D1548868

Sustainable Development

Sustainable Development

EXPLORING THE CONTRADICTIONS

Michael Redclift

London and New York

First published 1987 by Methuen & Co. Ltd.
Reprinted 1989 and 1991 by Routledge
11 New Fetter Lane, London EC4P 4EE

Simultaneously published in the USA and Canada
by Routledge
a division of Routledge, Chapman and Hall, Inc.
29 West 35th Street, New York, NY 10001

Printed in Great Britain by Richard Clay Ltd, Bungay, Suffolk

British Library Cataloguing in Publication Data
Redclift, Michael
  Sustainable development: exploring the
  contradictions.
  1.   Developing countries — Economic
  conditions.
  I.   Title
  330.9172'4   HC59.7

  ISBN 0-415-05085-5

Library of Congress Cataloging in Publication Data
Redclift, M.R.
  Sustainable development.

  Includes index.
  1.   Economic development — Environmental aspects.
  2.   Environmental policy.   I.   Title.
  HD75.6.R43   1987   363.7   87-5710

  ISBN 0-415-05085-5

Contents

Acknowledgements

Writing this book has been a voyage of intellectual discovery for me and, as with most voyages of this kind, I have benefited from meeting many interesting people *en route*. Among the people who have contributed ideas, reactions and comments, as well as considerable professional support, I must mention Colin Sage, Philip Lowe, Bryn Green, Jonathon Porritt, Osvaldo Sunkel, Juan Martinez-Alier, David Goodman, Richard Norgaard, Bernard Glaeser, David Pearce and Fred Buttel. None of them bear responsibility for the opinions expressed. Anne Weekes typed the manuscript through various drafts with constant good humour. I also wish to thank Eleanor Rivers of Methuen for her encouragement and enthusiasm.

Finally, I refuse to descend to the slipshod sexism, usually masquerading as 'professionalism', which leads many male authors to thank their spouses for providing unpaid material and emotional support services, without which their book would not have been written. I did *not* spend my weekends working in my study while my wife took the kids for long country walks. Nanneke Redclift did make a real intellectual contribution to this book and one that I cannot begin to repay. In this, and in sharing domestic and emotional support, I have her more than anyone to thank.

1

Introduction

For three months in early 1983 a massive forest fire destroyed over 3.5 million hectares on the island of Borneo (Indonesian Kalimantan). This charred area, nearly the size of Taiwan, included 800,000 hectares of primary, tropical forest and 1.4 million hectares of commercially logged woodland. An additional 750,000 hectares had been secondary-growth forest under shifting cultivation, and 550,000 hectares consisted of peat swamps. As E. C. Wolf (1985) argues, any area which gets five times as much rainfall as New York City or London should be difficult to ignite. However, human actions had paved the way for the biggest recorded 'natural' conflagration in history. The ranks of cultivators had risen in Kalimantan by many thousands, some of them settled as part of Indonesia's massive transmigration programme. Loggers promoted the fire's spread by leaving damaged trees standing after selective commercial logging. Researchers at the University of Hamburg suggested that changes in the turbidity of coastal waters, due to soil erosion in South-east Asia, may have altered regional atmospheric currents, contributing to the drought. As trees dropped their leaves in an effort to conserve moisture, the forest floor became a vast, tractless tinder-box. Like so many other 'natural' disasters the destruction on Kalimantan had human causes.

While the tropical forest on Kalimantan was being destroyed, work was commencing in Arizona on a sealed-off, glassed-in space capsule, nicknamed 'Biosphere II'. Perched in the Catalina mountains, 35 miles north-east of Tucson, this cubic glasshouse was designed to 'be as removed from the life-supporting systems of Earth as a space station' (Johnson 1986). It consists today of 17,000 square feet of plant-tissue-culture laboratories, plant and aquaculture greenhouses and a support building complex, the property of Space

Biosphere Ventures. In 1989 Biosphere II's airlock will be finally sealed and the glasshouse capsule will rely, as does our planet, on its own life-support mechanisms. The aim of Biosphere Ventures, we are told, is 'to give biological Earth a mate with which it can conduct a dialogue . . . it is a starting point for a dialogue of biospheres' (Johnson 1986, 22). It should also give us pause to think. The life-support systems which have enriched our planet and given us such a diversity of species are fragile, not only in tropical forests like those of Borneo, but also in English wetlands and chalk downs. The environment is frequently placed in jeopardy by development. At the same time, we are heavily involved in recreating nature, reassembling the parts and cloning the genes, in an attempt to remove ourselves from environmental constraints. We are, literally, 'producing' nature for the first time, while we are busily engaged in destroying it for the last time. This book is about both processes, about the destruction of life-support systems and their creation. It is about the meaning we attach to 'sustainable' development and the contradictions which sustainable development implies. Between Borneo and Arizona lies a great deal of human history, as well as geography. My purpose is to analyse that history, and its underlying momentum, in order to discover why development has taken the course that it has and what we can do about it.

Sustainable development seems assured of a place in the litany of development truisms, but to what extent does it express convergent, rather than divergent, intellectual traditions? The constant reference to 'sustainability' as a desirable objective has served to obscure the contradictions that 'development' implies for the environment. Instead of bringing intellectual rigour to the discussion of environment and development, we frequently encounter moral convictions as substitutes for thought. However important these convictions, in partnership with rigorous analysis, they are no substitute for it.

In exploring the relationship between development and the environment we will need to construct a model of how it has changed over time: a historical account of the environment and development. Equally important, we will need to make clear the international linkages which provide the transformation momentum behind environmental change. These international linkages involve the transfer of capital, labour and natural resources. In exploring sustainable development, we are necessarily concerned with all three: with capital and labour, as well as the 'natural' resources that human beings have 'naturalized' through their own efforts (Smith 1984).

In the pages that follow it will also be clear that the environment, whatever its geographic location, is socially constructed. The environment used by ramblers in the English Peak District, or hunters and gatherers in the Brazilian Amazon, is not merely *located* in different places; it means different things to those who use it. The environment is transformed by economic growth in a material sense but it is also continually transformed existentially, although we – the environment users – often remain unconscious of the fact. This book aims to deepen our understanding of environmental change as a social process, inextricably linked with the expansion and contraction of the world economic system.

In a previous discussion (Redclift 1984) it was argued that political economy and environmentalism each stood to gain from sharing an analytical perspective. The environmental 'crisis' in the South was the outcome of an economic, structural crisis. At the same time it was argued that the political economy of development needed to incorporate environmental concerns in a more systematic way. This book begins where that discussion ended: with an approach which it is hoped is more integrated in both disciplinary and historical terms. The reader who was enjoined to take 'cues from societies whose very existence "development" has always threatened' (1984, 130) will note that the cross-cultural and historical approach that follows is a logical consequence of considering development and the environment as an integrated process. It hardly needs to be said that we still have a long way to go before this conceptual integration is complete.

Since the late 1960s there has been considerable discussion of development, both as a concept and in concrete historical settings. Important differences exist between neo-classical and Marxist interpretations, and within each of these competing paradigms. In this book 'development' is regarded as an historical process which links the exploitation of resources in the more industrialized countries with those of the South. The perspective adopted is that of political economy, in which the outcome of economic forces is clearly related to the behaviour of social classes and the role of the state in accumulation. At the same time, it is central to the argument of this book that 'development' be subjected to redefinition, since it is impossible for accumulation to take place within the global economic system we have inherited without unacceptable environmental costs. Sustainable development, if it is to be an alternative to unsustainable development, should imply a break with the linear model of growth and accumulation that ultimately serves to undermine the planet's life support systems. Development is too closely associated in our minds with what has occurred in western capitalist societies in the past, and a handful of peripheral capitalist societies today. To appreciate the limitations of development as economic growth – the starting point for the discussion in chapter 2 – we need to look beyond the confines of industrialized societies in the North. We need to look at other cultures' concept of the environment and sustainability, in historical societies like that of Pre-Columbian America, and in the technologically 'primitive' societies which present-day development serves to undermine.

Environmental change and structural underdevelopment

We begin with a dubious legacy. The environment has suffered more neglect at the hands of social scientists than any comparable subject. If it has fallen to natural scientists to understand environmental change without recourse to the methods and analytical tools of social science, it is hardly our place, as social scientists, to criticize. In the absence of theoretically refined work on environmental issues most social scientists, when they have entered the field at all, have been content simply to collect data, provide criteria for land classification (land use) or for ways of 'costing' environmental impacts and losses (such as Environmental Impact Assessment). Social scientists have used ecological processes in a metaphorical or descriptive sense ('cultural ecology', 'urban ecology'). The environment has fallen between too many disciplinary stools and, as we shall see, thinking about the environment has become divorced from social and economic theory.

Nevertheless a minority of scholars has given significant attention to the economic modelling of environmental variables (Pearce 1985, Norgaard 1984b). Few authors within the Marxist tradition have attempted to integrate the environment within their theoretical framework; those who have made this attempt reveal the neglect of decades (Smith 1984, Blaikie 1985, Vitale 1983, Galtung 1985). Sociologists, in particular, have deserted the historical project inherited from both Weberian and Marxist directions. What Humphrey and Buttel (1982) describe as 'their collective celebration of Western social institutions' has caused them to regard 'energy-intensive industrial development (as) the natural end point of a universal process of social evolution and modernization'. Perhaps unsurprisingly, it is philosophers who appear to have found, most recently in 'deep ecology', a substantive research problem with which they feel comfortable (Sylvan 1985a & b, Devall and Sessions 1984, Naess 1973).

This book addresses itself to the neglect of the environment by most social scientists. As such it is partly the product of an increasingly challenging and provocative literature debate among those interested in both 'development' and 'conservation' (McNeely and Pitt 1985, Conway 1984, Goodland 1985, Blaikie 1985, Saint 1982, Norgaard 1984b). However, not for the first time theoretical discussion has failed to cross the North/South divide, and progress in regarding environmental problems as linked by the development process is slow. Reassessments of theory would benefit from a more systematic attempt to relate environmental change in the North to structural development processes in the South and vice versa. At the same time the objective is also to locate our conception of the 'environment' within a broader comparative framework, one which distinguishes the historical role of the environment within capitalist development. Finally, this analysis proceeds from an attempt to identify common elements in a political economy of the environment relating environmental change to 'superstructural' factors, such as ideology and policy, and at different levels of political complexity. The intention is to provide a structural analysis of the environment in which the development process illuminates environmental change in different societies at both a material and a phenomenological level.

The structure of the book reflects the elaboration of this argument, beginning with the discussion of the concept 'sustainable development' in chapter 2. In this chapter the ecological theory underpinning sustainability is reviewed, and the role of energy and population are assessed in relation to the environment.

Chapter 3, which follows, takes issue with the failure of conventional economics to provide an adequate theoretical account of environmental factors and compares the arguments for a more inclusive economics with the view that other paradigms may provide alternative insights and policy agendas. The perspectives of 'deep ecology' and orthodox Marxism are discussed, and their limitations revealed.

Chapters 4 and 5 examine the dimensions of the global environmental 'crisis' from the perspective of international political economy, arguing that the process of development cannot be divorced from the international economic system in a specific historical phase. It is international economic structures, as well as intellectual traditions, which impede our progress. Chapter 5 demonstrates how specific economic linkages have evolved between North and South, helping to establish environmental conditions for development and accompanying problems for developing countries. This chapter then examines the explicit recognition that sustainability must be linked to a new 'style' of development, which has been convincingly argued by the Santiago office of the United Nations Environment Programme (UNEP).

Chapter 6 returns to what is happening on the ground, by examining the relation between commodity production under capitalism and the kind of environmental transformation to which commodity production and the market give rise. A detailed case study of Eastern Bolivia reveals the extent to which development solutions that are ecologically and agronomically sustainable confront both structural obstacles in the wider economic system and the conflicting effects of poor peoples' strategies to survive.

Chapter 7 takes a closer look at the 'environmental management' approach which is directed at resolving, or reducing, the contradictions exposed by the development process. It discusses the relevance of environmental management and conservation in developed countries for the experiences of developing countries and argues that the experiences of indigenous peoples in 'managing' their environments should be an essential element in a more relevant approach. The organization of poor people around political struggles for environmental objectives is also discussed, with examples drawn from India, East Africa and Mexico.

Chapter 8 explores the 'frontiers of sustainability' by linking two intellectual traditions of scientific thought. The potential strengths and weaknesses of Marxist approaches to the environment, which view the environment as a 'commodity' under capitalism, are linked to the 'production of nature' via biotechnology and genetic engineering. The weakness of Marxist theory in failing to integrate reproduction and ecological sustainability in its account of the development process is paralleled by the failure in the reproductive and biotechnological sciences to address the social and economic implications of transforming nature. Finally, in the conclusion, the argument is restated and re-examined, both for its intellectual coherence and its implications for future practice. If sustainable development is founded on contradictions, how should we seek to resolve them in practical policy terms?

Intellectual ancestry

Much of the environmental debate has been conducted with only fleeting references to the development of capitalism, the process which assumes greatest explanatory power in this book. However a number of different theoretical currents in this debate can be distinguished.

One current of opinion has its roots in Herbert Spencer and has sought to explain human behaviour as the 'internalization' of nature. The premise is that a biological basis exists for social action and behaviour: biological determinism. This approach seeks to explain social institutions like property in terms of their biological 'roots' and nationalism in terms of territoriality. But the writing of Robert Ardrey and Desmond Morris is illustrative of a broader and more academically respectable perspective: which includes ethology and, most recently, sociobiology. For sociologists whose training has taught them to distance themselves from the evolutionary perspectives derived from nineteenth-century natural science, such perspectives are irrelevant if not potentially dangerous (Buttel 1983).

Social theory and the environment

What can be detected in the historical development of the social sciences is not so much a conclusive rejection of organically based theory as a continuing tension between two scientific traditions. Some disciplines, notably geography and anthropology, have accommodated both traditions to some degree; others, like sociology and economics, have proceeded by exorcizing the ghost of 'organic' theory at some cost to their own paradigms. Two examples of this ambivalence towards theoretical positions which incorporate models derived from the natural sciences are the experiences of 'human ecology' in urban studies (Park and Burgess 1921) and of 'cultural ecology' within social anthropology (Geertz 1963, W. Wolf 1959). Both theoretical positions were associated with academic 'schools' within social science disciplines, but neither succeeded in building social science theory around human/natural environment interaction.

To understand fully the limits placed on our view of 'development' by the separation between natural and social sciences, we can begin with the Founding Fathers of sociology: Marx, Weber and Durkheim. The development of specifically 'social' theory in the late nineteenth century was partly an attempt to fill the void left by the development of economics, on the one hand, and the unsatisfactory eighteenth-century legacy of biologically related social theory on the other. Marx, for his part, inveighed against the emphasis on scarcity in Malthusian thought and developed Ricardian political economy in a way that was distinctly optimistic. For the first time since the Renaissance human capacities were viewed as more than a match for nature. Commodity production under capitalism served to energize the productive system. The environment performed an enabling function, but it was impossible to conceive of 'natural' limits to the material productive forces of society. Indeed the barriers that existed to the full realization of resource potential were imposed by property relations and legal obligations rather than resource endowments. They were social barriers.

Max Weber also sought to dispose of organically conceived theories in a way that has left its mark on contemporary sociology. Civilizations were regarded in their own right as unique and enduring cultural traditions. While emphasizing an historical dimension, something that informed all his work, Weber insisted on the distinctiveness of social processes against a simplistic evolutionary perspective that viewed all societies as passing through successive stages. To some extent Weberian sociology has been distorted in the retelling, and we are familiar in developing countries today with the unilinear account of modernization which takes Weber as its point of departure. However, Weber himself was at pains to establish a synchronic view of social development, both in opposition to that of historical materialism and to the Comtian positivism which had so influenced European thought in the previous generation.

The third founding father of sociology, Durkheim, was equally determined to break free of biologically grounded social theory. For Durkheim the search for social explanations for social phenomena was a methodological posture, and the development of sociology could be measured in terms of its attainment. Against Spencer and Social Darwinism, Durkheim insisted that to derive social processes from the natural world was actually prejudicial to understanding. In particular the operation of the division of labour in society, although it had analogies in nature (organic solidarity), was essentially the outcome of a technological process of differentiation. *Just as the Weberian tradition emphasized the individual's contribution to social action, and the Marxist tradition emphasized the capacity of social action to transcend the individual, so the Durkheimian tradition asserted the primacy of the social, even the collective, mind.* In the light of these intellectual precedents it is not surprising that 'an implicit taboo [exists] against incorporating ecological variables in sociological analysis' (Buttel 1983, 11). It is clear that, by the early part of this century, the social sciences had incorporated within their view of development two features of continuing importance. First, the notion that economic growth was essential to the development of social institutions and was made possible by exploiting, rather than seeking to conserve, natural resources. Second, although theoretical models might draw analogies with natural systems, explanatory theory was largely taken up with finding non-naturalistic causes for progress in human societies.

One intellectual approach, which has proved to be more insistent and more capable of generating widespread intellectual support during recent years, can be described as 'Neo-Malthusian'. This perspective returns to challenge the view that, since the late nineteenth century, biological and evolutionist accounts of development lack credibility. In essence Neo-Malthusianism rests on the Malthusian principle that population cannot exceed resources without famine or disease providing natural checks on population growth. In the view of Neo-Malthusians recent successes in reducing mortality rates, especially dramatic in many Third World countries, have given added importance to the Malthusian edict. In the influential 'tragedy of the Commons' discussion Hardin (1968) argued that people are incapable of putting 'collective' interests before 'private' ones. Hence the resource base was constantly under threat from behaviour which, at a disaggregated level, was logical. In later writing Hardin and others have argued that pre-emptive, even coercive, action is needed to control population and conserve resources. As technological society progresses, the scarcity-induced control mechanisms which formed an important part of Malthus's argument fail to work as expected. One dimension that has been explored by Commoner (1971), Ehrlich (1974) and Myers (1979) is the political impasse. Social and political institutions change too slowly, and are unable to accommodate themselves to the realities of new resource pressures. The problem of population increase continues to be used by critics to undermine environmentalist positions.

There are a number of objections to the Neo-Malthusian position and its variants. Marx's original strictures about Malthus's writing are still one source of criticism. The Neo-Malthusians do not address distributive issues with the urgency they require. Marx put it more bluntly, arguing that the Malthusians of his own time emphasized the 'limits of nature' for ideological reasons – it justified them in the view that nothing could be done about poverty.

Neo-Malthusianism also meets objections from a geopolitical standpoint. From the perspective of a less developed country the emphasis on population and 'global' solutions looks suspiciously like an attempt to evade the issue of the role of international economy in structural underdevelopment. The developed countries have an interest, it is claimed, in drawing attention to resource scarcities, since they imperil *their* economic development. They have much less interest in a fundamental restructuring of the international economy which might relieve many of the resource pressures experienced by societies in the South. Hence the very enthusiasm for environmental issues in countries like the United States sometimes creates intense suspicion in the South.

Another, equally forceful, approach to population and resources corresponds with O'Riordan's (1981) 'ecocentric' category. Deriving advantage from the 'limits to growth' discussion of the 1970s (Meadows *et al*. 1972), it is argued that the problem is *not* the balance between population and resources but the *ends* to which resources are put in the pursuit of economic growth. In the process of 'development', it is argued, we in the industrialized countries, have lost our 'respect' for Nature, and with it our margin of freedom to proceed by trial and error (Dasmann 1975, 19). The ecocentric perspective takes issue with the objectives of development as well as the means. In this respect, at least, it represents a more radical break with orthodoxy than other ideological/paradigmatic positions.

One of the features of the last few years is the way in which the ecocentric approach has become much more concerned with the structural relationship between both developed and less developed countries. Within a radical 'Green' perspective the implications of radical action must, ultimately, bear on 'us' as well as 'them'. International food policy and international trade increase the economic dependence of less developed countries and reduce the sustainability of their environments. Perhaps the question we need to pose is that raised in the British response to the World Conservation Strategy (1983): what does sustainability imply for *our* development?

The political economy of the environment has, necessarily perhaps, become increasingly heterogeneous. In less developed countries, 'ecocentric' positions seek to rediscover what existed before the colonial embrace distorted their development process (Gonzalez 1979). This implies building bridges to their own past. In developed countries the emphasis is rather on the growing malaise within post-industrial society, and what this implies for work, leisure and culture (Gorz 1980, Bahro 1982, Williams 1984).

Neither Neo-Malthusianism nor the ecocentric perspective give much emphasis to the way in which capitalist development makes use of the environment. The ecological crisis is depicted as larger than politics, larger even than capitalism – tempting some towards a position which Pepper (1984) sees as 'ecofascism'. Certainly both approaches disavow conflict in the Marxian sense, as a historical and necessary source of change and liberation. The appeal to balance, to good husbandry, to the defence of the species, appears to put Nature before People; but it does so in a way that reduces the role of human beings in their own development. There are objections to the market as a principle of economic organization, notably from the ecocentric perspective, but little interest in exploring how the market works comparatively. The commitment to stable-state resource allocation, and to a zero-growth position, in which use values are substituted for exchange values, precedes any systematic attempt to establish how these new goals can be legitimized or brought nearer under capitalism. Sustainable development is the objective of many perspectives on the environment, but the role of the market in defining the various outcomes is considered in few of them.

The environment and capitalist development

This book concentrates on the way that environmental issues are socially constructed under capitalism because, in general, little consideration has been given to the impact of capitalist development on the environment in any of the perspectives discussed. Even the approach to the environment which probably can count upon most support – consensus management of the environment or environmental planning – is mainly concerned with objectives that appear to lie outside the camp of market economics, such as conservation and physical planning.

The structural linkages which exist between economic development and the environment in the North and South (policies such as the disposal of food-grain surpluses from the North (Public Law 480 (PL480)) in the past; EEC food 'mountains' in the present) radically affect the environment in the South. The penetration of the South by new agricultural production technologies, marketing and contract farming, have also served to shift agriculture in parts of Latin America and Africa away from traditional, environmentally sustainable systems towards greater specialization and economic dependency. These problems are more acute when so many countries in Africa and Latin America have enormous external debts which they are urged to repay by more specialized exports of cash crops, forest products, etc. Changes in the environments of the South need to be understood, then, in terms of the international redivision of labour. It also makes it imperative that we consider what has been lost, as well as gained, in the development process. As E. Wolf (1982) demonstrates in *Europe and the People without History*, the emergence of European colonialism and, at a later stage, industrial capitalism served to obscure the history of cultures with which contact was made. Nevertheless 'capitalism did not always abrogate other modes of production, but it reached and transformed people's lives from a distance as often as it did so directly' (p. 311). When the transformation took place at a distance, sustainable resource use was sometimes practised, despite the exigencies of the market. Sustainability of the environment in these pre-industrial societies was not divorced from traditional agricultural practice; it had no independent ontological status.

The World Commission on Environment and Development (Brundtland)

Increasing concern with environmental problems in developing countries and the failure to relate these problems to development issues led to the establishment of the United Nations Commission on

Environment and Development in November 1983. This Commission, under the energetic leadership of Norway's prime minister, Mrs Brundtland, consisted of twenty two people from both developed and developing countries. The objective of the Commission, according to its interim statements, was to focus on the causes of environmental problems rather than the effects of environmental degradation. Unlike earlier international reports the Brundtland Commission did not wish to report on trends in the world environment, since so many reports existed already. The main objective was to undertake public hearings in various countries, at which members of the public and community leaders could give evidence about the relationship between development and the environment and the Commission could visit selected sites.

The members of the Commission, for their part, were not chosen for their expertise as environmental 'specialists', but as prominent people who were appraised of the facts and were prepared to ask relevant questions about the causes of environmental problems. Mrs Brundtland, in a radio interview, argued that the Commission was stronger for including both an American Republican and several more left-wing members from developing countries who might be prepared to support similar measures on development and the environment. In the present economic climate, it was felt, when international co-operation on development issues had deteriorated, it was necessary to provide an opportunity to regroup and press the case for sustainable development. Mrs Brundtland has expressed publicly her conviction that the free-market principles which are often thought to govern international economic relations are inappropriate and indeed prejudicial to the interests of better environmental management in the South.

In some important respects the Brundtland Commission was a significant advance on previous global exercises of a similar kind. In a series of public hearings during 1985 and 1986 in Oslo, Jakarta, São Paulo and Harare, the Commission met to hear evidence and visited areas which have been severely affected by environmental problems. Their mandate was to 'formulate innovative, concrete and realistic action proposals . . . to assess and propose new forms of co-operation . . . and to raise the level of understanding and commitment' (Brundtland 1985a, 9). Explicit in most of the Commission's preliminary documentation is the fact that change can only come about as a result of political action, and that the environment is a heavily contested area, however much consensus may seem to surround the subject.

In the original documents the Commission draws attention to what it calls a 'standard agenda' of environmental concern which it wants to call into question. This agenda commits a number of errors of bias or omission which the Commission seeks to correct. First, it is usually the effects of environmental problems that are addressed in public documents, as we have seen. Second, environmental issues are usually separated from development issues and frequently pigeon-holed under 'conservation'. Third, the Commission complains that critical issues, such as acid rain or pollution, are usually discussed in isolation, rather as if solutions to these problems can be found in discrete areas of policy. Fourth, the Commission criticizes what it sees as a narrow view of environmental policy, which relegates the 'environment' to a secondary status – it is 'added on' to other, more important development issues.

In its refusal to accept that environmental problems can be addressed through their causes, and in its critique of conventional environmental management, as practised in the developed countries, the Brundtland Commission is expressing views similar to those expressed in this book. In many ways the criticism of the 'standard agenda' and the global interpretation that the Commission seems prepared to advocate leave the impression that Brundtland is much the most radical departure we have seen. When the full document is published in early 1987 it will be worth serious attention, not only because of the evidence it is likely to provide of the links between poverty and the environment in developing countries, but also because it is a mark of the seriousness of the problem that a group of mainstream political leaders should have helped to put such a document together. It remains unlikely, however, that the developed countries (or even the developing ones) will put into action the measures advocated by the Brundtland Commission. It is the argument of this book that they cannot do so without involving themselves in very radical structural reform, not only of methodologies for costing forest losses or soil erosion, but of the international economic system itself. However, before reviewing the process through which the environment has become internationalized, transformed and ultimately either 'managed' or artificially created in laboratories, we need to address the central question: what is 'sustainable development'?

2

Sustainable development: the concept

Development is usually defined principally in terms of economic growth: as countries experience increased growth their productive capacity expands and they 'develop'. As long as population increases – and there are few contemporary societies in which population is *not* increasing – then it is difficult to imagine development without economic growth. The problem of containing the negative effects of economic growth on the environment cannot be reduced to demographic factors alone. In chapter 4 some of the negative implications of pursuing economic growth in the South are examined, since they frequently call into question 'development' itself. For the moment we need to consider the immediate question: to what extent is economic growth an adequate measure of development?

The crudest, and most familiar, indicator of development is gross national product (GNP). The limitations of GNP as a measure of development are easily identified. First, GNP measures 'productive' activity in a very narrow way, excluding, for example, the productive activities of the household because many of these are undertaken by women and children. It is a measure of 'formal' sector activity, whether in the primary sector (such as agriculture), or in manufacturing and services. The 'informal sector', in which markets exist but are not fully reported statistically, and in which people produce for their own consumption, is not represented in GNP figures. These informal activities are particularly important when we consider the environment in the South: collecting firewood, cooking food, feeding, clothing and housing people. None of these activities are adequately represented in GNP statistics.

In addition, GNP is a very blunt instrument for measuring economic development without considerable attention being given to demographic profiles. Per capita figures for economic growth, for example, disguise the number of dependants within families, the number of single parents and elderly people without dependants. Since much productive activity takes place in the home, and between households in communities, per capita figures tell us very little about the relationship between income, wealth and patterns of income distribution even among people of the same class.

Economic growth measured through GNP is also an inadequate measure of how production is deployed. All measurable production activity is considered the same, whether it is channelled towards arms expenditure or the maintenance of a primary health-care system. This makes it impossible to distinguish between countries which spend a high proportion of their income on defending themselves, such as Israel and the Soviet Union, and those which have no army, like Costa Rica.

GNP figures also fail to distinguish between groups of people, especially social classes, within a country. Some poor countries share their wealth much more equally than others with similar GNP standing: in Latin America one can cite Cuba and Brazil as examples of relatively equitable and highly inequitable distribution. Clearly a measure of development which does not consider the distribution of income and wealth is hardly satisfactory. Societies may be both 'developed' and 'underdeveloped' at the same time, most countries in the South are, but the extent to which wealth is geographically and socially concentrated needs explict attention.

Finally, GNP statistics record the productive utilization of resources, whether or not these resources are renewable. Moreover, if productive activity is associated with the costs of economic growth, through pollution control, for example, this is also entered under GNP. Deforestation, bringing with it a loss of resources, is usually treated, for example, as a net contributor to capital growth (Pearce 1986). From an environmental standpoint, then, GNP is a particularly inadequate guide to development since it treats sustainable and unsustainable production alike and compounds the error by including the costs of unsustainable economic activity on the credit side, while largely ignoring processes of recycling and energy conversion which do not lead to the production of goods or marketable services. As we shall see later 'sustainable' resource uses play a particularly important role in the transformation of economic activity in the South.

The use of other social and economic indicators represents an advance on the crude measurement of GNP. The World Bank's annual *World Development Report*, for example, makes use of a number of such indicators: average annual rates of inflation, adult literacy, life expectancy at birth, average index of food production and others. In addition the structure of productive activity, or that part of it amenable to official statistics, receives attention in reports like those of the World Bank. Considerations include the sectoral 'divisions' of production (primary, secondary, tertiary), the growth of investment, the structure of demand (consumer goods/savings/public goods) and the production and consumption of energy. Clearly a definition of sustainable development needs to take account of the wide variations in the industrial and productive structures of different countries. 'Development' in the United States, as its economy is currently organized, requires 370 times as much energy per capita as it does in Sri Lanka. Does this imply that the United States cannot achieve sustainability, given its economic structure? Or that Bangladesh cannot achieve 'development', given its economic structure? As we shall see, sustainable development is usually thought of in the context of developing countries' sustainability, without attention being given to the international structures within which such countries are located.

Ecological systems and agricultural development

As Dasmann (1985) observes, reference to the sustainable uses of land and biotic resources within ecology has several antecedents, in forestry and wildlife management in particular. Within plant ecology the key concept was that of successional change in plant communities (Clements 1916). Successional change provided a model, drawn from nature, for the management of forests and rangelands. Forests could recover through natural processes from regular cutting and burning. Animal populations could also re-establish themselves after being hunted almost to extinction, provided that their natural habitat was maintained. The evidence that species and natural communities might not recover from excessive destruction of their habitat increased during the 1950s (Ellison 1954, Burcham 1957). More recently the recognition that we are allowing the pursuit of agricultural growth seriously to damage ecological sustainability in developed countries has attracted considerable attention (Shoard 1980, Green 1981).

The most important feature of the dynamics of ecosystems is that 'evolutionary adjustment' (Odum 1971, 35) to new patterns of natural resource use takes considerable time. The homeostatic controls that exist within natural communities, and that enable them to achieve succession, are only effective if these ecosystems are protected from rapid change. Ecological succession typically culminates in a climax system of high diversity, large biomass and high stability (Bartelmus 1986, 44). However, mature ecosystems displaying these characteristics, such as the tropical forests, achieve stability through shifts of energy flows away from production and towards the maintenance of the system itself. As Bartelmus (1986, 44) explains: 'In young systems the rate of gross production of biomass and organic matter tends to exceed the rate of community respiration, that is the maintenance costs of the ecosystem. Mature systems on the other hand exhibit equal or near-equal rates of production and respiration.'

It is not difficult to perceive where the interest of human groups lies – in the maintenance of young, highly productive ecosystems, in which organic matter and biomass are not allowed to accumulate. Maximizing agricultural production inevitably leads to the removal of mature ecosystems or steps to prevent their developing, at the cost of confounding nature's strategy of maximum protection or adaptation.

In addition, the maintenance of an ecosystem in an artificially young state of high productivity, under crop production for example, requires enormous energy subsidies in a form not available in nature. Fertilizers, fuels for machinery, irrigation technology, genetic selection of species and pest control, are all facets of this attempt to renew immature ecological systems in a state of high productivity. In essence agricultural development implies a necessary threat to ecological succession, in which costly energy subsidies replace natural processes. Sustainability, in this primary sense, is not only endangered by ecologically unwise agricultural practices, it is endangered by all agriculture. However, the level of conversion to ecologically harmful practices is such that the problem we are facing is not simply how to compensate for the interruption of ecological succession, it is frequently how to ensure that production itself does not degrade resources beyond the point of renewal.

The effects of agricultural development on the capacity of the ecosystem to achieve high levels of renewal has been discussed by G.

Conway (1984, 1985a). Conway refers to what he calls the four properties of agro-ecosystems – productivity, stability, sustainability and equitability. He points out that these properties are relatively easy to define, but much less easy to measure. Productivity is the yield or net income per unit of resource. Stability is 'the degree to which productivity is constant in the face of small disturbances caused by the normal fluctuations of climate and other environmental variables' (1985a, 35). Sustainability refers to the system's ability to maintain productivity in the face of a major disturbance, such as that caused by soil erosion, farmer indebtedness, an unanticipated drought or a new pest. The loss of sustainability is then expressed through declining productivity or a sudden collapse in the system. Finally, equitability expresses the distributive aspects of the agro-ecosystems: '. . . the more equitable the system the more evenly are the agricultural products shared among the members of, say, a farm household or a village' (Conway 1985a, 35).

Table 2.1 Agricultural development as a function of agro-ecosystem properties

	Productivity	Stability	Sustainability	Equitability
Swidden cultivation	Low	Low	High	High
Traditional cropping system	Medium	Medium	High	Medium
Improved	High	Low	Low	Low
Improved	High	High	Low	Medium
? Ideal (best land)	High	Medium	High	High
? Ideal (marginal land)	Medium	High	High	High

Conway argues that these properties of an agro-ecosystem can also be regarded as indicators of the performance of that system. Thus traditional agricultural systems, such as shifting cultivation (swidden), are generally low in productivity and stability, but high in equitability and sustainability (see table 2.1). Traditional but sedentary cropping systems tend to be more productive and stable but retain a high degree of sustainability. They are also relatively equitable. The introduction of new technology, such as the high-yielding rice varieties associated with the Green Revolution, greatly increases the productivity of the system but exposes the system to other hazards, notably attack from pests and diseases. More recent improved varieties have served to improve the stability of the agro-ecosystem without losing its high productivity, but sustainability is still low, largely because of the extensive use of chemical inputs to the system. Conway argues that the 'ideal' solution lies in improving equitability, at the cost of sacrificing some of the productivity associated with the Green Revolution technology (Conway 1985a, 36–7). This conceptual framework has been utilized in several different geographical settings, both as an adjunct to farming-systems research and as a means of effecting 'rapid rural appraisal' (Conway 1985b and c). The wider implications of looking at cropping and other agricultural systems, together with their energy requirements, are discussed later in this chapter.

As Dasmann (1985) has observed, the concept of sustainability received its greatest boost from the publication of the World Conservation Strategy (IUCN 1980). The World Conservation Strategy (WCS) (IUCN 1980) and a series of related documents (UNEP 1981, South Pacific Commission 1980) brought the concept of sustain-ability to the attention of a much wider audience. In addition, the Strategy explicitly linked the maintenance of ecological processes and life-support systems, the first of its three programme priorities, to the sustainable utilization of resources and the maintenance of genetic diversity, the other two priorities which were advanced in the report.

The World Conservation Strategy argued that the maintenance of ecological processes could only be brought about if urgent considera-tion were given to three specific conservation objectives: the utilization of good cropland for crops rather than cattle-raising, the ecologically sound management of crops and the protection of watershed forests. At the time of writing none of these objectives are assured. Crop production often took place on marginal land which was ecologically fragile. Similarly, floodplain agriculture was not managed in ways that preserved ecological processes. The disruption of traditional agricultural patterns had seriously damaged soils and had other damaging effects on the ecosystem – for example, insecticides destroyed fish stocks in paddy fields, bringing serious dietary consequences. The financial burden of heavy fertilizer use, especially in countries without their own oil supplies, needed to be offset by better use of organic wastes. Recycled organic materials and shorter fallow periods, substituting mixed for single cropping systems, were necessary if agricultural development was to be maintained in the face of population increases (IUCN 1980).

Similarly, the World Conservation Strategy made a valuable contribution to the discussion of development by pointing out that maintaining genetic diversity was not merely ecologically necessary, it was necessary to the development of agriculture itself. The extinction of genetic varieties, vividly portrayed for the humid tropics by Norman Myers (1979), reduced the gene pool and threatened the ability of human beings to make adaptation to changing ecological conditions. Maintaining genetic diversity did not necessarily mean on-site preservation in the wild. It also meant off-site preservation, especially in forestry and fisheries which, unlike crops, had not proceeded to domestication. Some plant species, such as *guayule* in Mexico's semi-arid regions, had a considerable potential for development which was rarely recognized.

Finally, the World Conservation Strategy identified, although it did not elaborate upon, the relationship between the productive capacities of natural resources and their human exploitation. The interest of human beings could not, ultimately, be divorced from that of the species which they utilized, since natural species were located in food chains and these food chains served the interests of human populations. In the view of the World Conservation Strategy, subsistence communities needed to be better equipped to utilize resources in a sustainable way: more attention needed to be given to the management of ecosystems, especially agro-ecosystems, by the people immediately dependent on these environmental resources.

In many respects, then, the publication of the World Conservation Strategy marked an important watershed in thinking about the environment and development. This should not blind us, however, to some of the deficiencies of this approach. The political and economic forces behind unsustainable practices received very little attention in the World Conservation Strategy documentation. Timber concessionaries are locked into powerful vested interests in less developed countries. Colonization, as harmful to many colonists as it is to the environment they depend upon, is favoured by governments as a 'painless' alternative to agrarian reform. The humid tropics are frequently opened up on the pretext of meeting food needs, although their effect is usually to jeopardize programmes for genuine increases in staple food production (Ewell and Poleman

South-North
Connection
exploitation

1980). As we shall see in the next chapter, the problems of initiating sustainable development alternatives are frequently undermined by the pursuit of illusory, and detrimental policies, whose origin lies in the North and in the relationship that is maintained between North and South.

Energy efficiency and agricultural development

The search for more sustainable development necessarily involves two interrelated dimensions. First, we need to consider to what extent we use energy efficiently within agriculture at the present time, since the development of more sustainable options may depend critically upon making better use of the resources we already command. Second, we need to consider population, together with ecological sustainability and energy efficiency, since the prospect of a decline in fertility in most parts of the South provides an incentive for more sustainable agricultural practices.

The development of a new agricultural chemistry in the 1840s by Boussingault in France and by Liebig in Germany alerted the world to the possibility of altering the balance between the sources of energy entering agriculture and those emerging from agricultural production. Liebig's idea was 'to change from an agriculture of spoilation to an agriculture of restitution' (Martinez-Alier 1985) by substituting chemical compounds, produced in the laboratory, for those that existed in nature. From an analysis of Peruvian *guano*, agricultural chemistry opened the way towards the commercial production of nitrogen and potassium-based fertilizers. The pioneers believed, correctly, that it would be possible to manufacture chemical fertilizers which possessed virtues similar to those of guano and other manure and, in so doing, they could raise the productivity of European farmers who were under pressure from rising urban populations to produce more food. By the end of the last century nitrogen fertilizers were being produced in Norway by hydro-electricity. Today they are at the forefront of the revolution in chemical/biological technology, the first and most dramatic stage known as the 'Green Revolution'. In the developed countries the simultaneous increase in agricultural production (through increases in mechanical power combined with chemical fertilizers) and the decrease in the size of the economically active agricultural population enabled a model to be erected of 'agricultural modernization' which was held to contain lessons for the development of poorer countries.

Table 2.2 sets out the principal variables in this model of agricultural modernization: the proportion of the population in agriculture, the number of tractors employed in place of agricultural workers, and the dependence on chemical inputs into agriculture. It can be seen that the countries with the most 'developed' agricultures, like the United States and the Netherlands, have reduced the size of their agricultural labour force by making intensive use of either mechanization or chemical fertilizers. Mechanical traction is more labour-displacing in its effects than chemical inputs, but the majority of farmers in many less developed countries make relatively little use of either type of technology. On the other hand, some developed or developing countries (such as Japan and South Korea) have developed their agricultural sectors principally around chemical/biological technologies, linked to some extent to small-scale mechanization. The important point about the figures in table 2.2 is that they demonstrate that there are several combinations of mechanization and chemical/biological technologies that are available and capable of being used in different societies. There is no one path to agricultural 'modernization'.

Table 2.2 Indicators of agricultural modernization (1980)

	Workforce in agriculture (%)	Tractors per 1000 agricultural workers	Chemical fertilizers (kg/ha)
Argentina	13.0	151	1.4
Brazil	38.2	21	11.5
Colombia	27.4	12	25.5
Chile	18.5	30	8.9
Venezuela	18.0	45	24.6
Mexico	36.0	16	32.4
Costa Rica	35.1	23	83.7
El Salvador	50.4	3	118.6
China (inc. Taiwan)	59.8	3	92.6
South Korea	38.6	1	209.1
Japan	11.0	171	147.3
United States	2.2	1934	51.6
Netherlands	5.4	576	514.3

Source: FAO 1980a and b

The different effects of agricultural technologies are shown in table 2.3. The points to bear in mind are as follows. First, in few cases are technologies used on their own; their combined effect is thus what matters. Second, the factors which determine the combination of technologies adopted are several; the agrarian structure of an area, the available natural resources and the links which exist with international sources of technological diffusion are among the most important. Clearly environmental considerations could play a major part in the choice of a technological option, but they are not the principal consideration. Instead, as Hosier *et al.* (1982, 180) insist, the key element is constraint rather than choice 'because constraints are the material coordinates within which choices can be taken'. Farmers necessarily use the kinds of technology that are made available locally and in a way they can afford, but that may not be those which are most appropriate for the environmental conditions under which they work. This does not mean, of course, that they necessarily use technologies which are environmentally damaging.

Historically, economic development has been linked to a progressive increase in energy consumption, and this is nowhere more

Table 2.3 The effects of different types of agricultural technology

Technology	Increase in land productivity	Increase in labour productivity	Increase in employment
Mechanical	+	+++	—
Genetic-biological	+	+	++
Chemical	+++	++	
Manual	+++	++	+

Notes:
1 Most technologies are used in combination, not on their own.
2 Technological options (combinations) are chosen to meet specific development objectives.
3 Environmental considerations rarely play a major part in the choice of a technological option.
4 Environmental effects are often complex e.g. (i) genetic-biological technologies *on their own* tend to accelerate soil exhaustion but also (via biotechnology) provide opportunities for less wasteful resource use (e.g. recycling); (ii) chemical technology has a negative effect through eliminating natural pest control but positive through (short-term) increases in soil fertility.
5 This analysis needs to be extended to *specific* resource uses, e.g. water, soils and woodlands.
Source: Gligo 1985

apparent than in the case of agricultural development. It should not surprise us, then, that underdeveloped countries consume much less energy per capita than developed countries. Ethiopia, for example, with a per capita income of US$120 per annum, shows a per capita energy consumption of the (coal) equivalent of 20 kg. At the same time Sweden, a representative developed country, shows a corresponding figure for per capita energy consumption of 6000 (coal equivalent) kg. Countries which are industrializing rapidly like Taiwan have quadrupled their per capita energy consumption in under 20 years (Hosier *et al.* 1982, 180).

Another useful example is that of Spain. In the 1940s and early 1950s Spanish agriculture provided employment for over half the population, although much of it was on a seasonal or part-time basis. In this period traction was still largely human or animal, and animal dung was the major source of fertilizer. As Martinez-Alier states, Spanish agriculture was 'technically more similar to Chinese agriculture than to the agriculture of North Atlantic countries' (Martinez-Alier 1985, 23). Indeed it was not until the 1970s that the number of tractors exceeded the number of mules in Spain. By the late 1970s the number of tractors had grown from 10,000 to over 400,000, the active agricultural population of two and a half million was half what it had been 30 years earlier, and the number of draught animals had decreased from 3.2 million to 1.1 million. The conclusion which Martinez-Alier (1985, 26) draws from an analysis of these, and other, figures is that:

> Not counting solar energy, the energy input into agriculture has increased more than production. While in 1950–51 one calorie of a 'modern' type of energy would help to 'produce' six calories of vegetable production, the ratio would be in the late 1970s down to one calorie per calorie . . . There has been an increase of productivity of labour, and a decrease in the efficiency of Spanish agriculture, in terms of conversion of 'modern' energy inputs.

This conversion process, through which energy finds its way into agricultural production, can be depicted in terms of alternative energy 'pathways' (figure 2.1). Drawing on the work of Simmons and Odum this diagram enables us to make some notional calculation of the population which can be fed from a given quantity of land. In those developed countries which have made the transition

Figure 2.1 Energy pathways in agriculture

Energy pathways in peasant agriculture

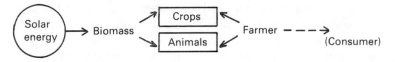

Energy pathways in capitalist agriculture

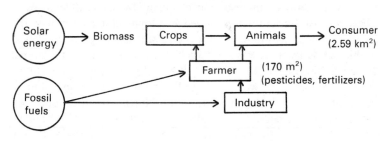

Note:
The combination of solar-energy and fossil-energy sources with crop/animal protein conversion enables each square metre (m^2) of land occupied by the farming population to support 32× this population in urban areas (2.59 km^2).
Sources: Simmons 1974, 194; Odum 1971

to a capital-intensive agriculture, including countries like Spain which have made it relatively recently, the conversion of energy has been the principal means through which food production has kept ahead of population, and the size of agricultural population has been reduced. Crops have been fed to animals, rather than directly to humans. Fossil fuels, both as requirements of mechanical traction and high-technology biochemical crop production, have been used to accelerate the production of agricultural goods, and to transform this production through food-processing and marketing.

Clearly, even setting aside the question of ecological systems and their sustainability, modernized agriculture is a very expensive consumer of energy, much of it non-renewable. Most comparisons between 'peasant' and modern capitalist agriculture are based on calculations of the productivity of land or labour, but they rarely include consideration of the energy efficiency of the agricultural system. Table 2.4 below provides a comparison of the energy efficiency of different agricultural systems by comparing the cost of energy inputs with the system's energy production. The energy efficiency of the system is determined by dividing energy production per hectare, per *annum* by the total energy input from all sources (oil, biomass, animal power, etc.).

The table shows that unmechanized systems of production which make use of animals for draught power (items 3 and 4), or which, like pastoralism in Africa, depend upon animal production as a final product, are relatively efficient users of energy. At the same time these systems are relatively autonomous, making little use of commercial sources of energy. Agricultural systems which show the

Table 2.4 Energy inputs, energy production and energy efficiency of agricultural systems (kcal/ha/annum)

Agricultural system	Total energy input (A)	Energy production (B)	Energy efficiency (B/A)
1 Pastoralism (Africa)	5,150	49,500	9.6
2 Crop rotation and fallow (Mexico)	675,700	6,843,000	10.1
3 Estate crop production (Mexico)	979,400	3,331,230	3.4
4 Estate crop production (India)	2,837,760	2,709,300	0.9
5 Maize (USA)	1,173,204	3,306,744	2.8
6 Wheat (USA)	4,796,481	8,428,200	1.7
7 Rice: irrigated (USA)	14,586,315	21,039,480	1.4
8 Apples (USA)	18,000,000	9,600,000	0.5
9 Spinach (USA)	12,800,000	2,900,000	0.23
10 Tomatoes (USA)	16,000,000	9,900,000	0.61

Notes:
(a) With unmechanized production, the dependence on animals for draught power (3 and 4) or as a final product (1) makes for relatively expensive energy conversion. However, pastoral systems (1) show only a limited capacity to produce energy per unit of land.
(b) The relationship between a *high level of energy consumption* and the *inefficiency of energy use* is marked (8, 9 and 10).
(c) Although there is a high correlation between the level of energy subsidy and yields, some low-input systems (2) are relatively high yielding.
(d) These calculations take no account of the use which different agricultural systems make of the effects of energy use on the recuperation of the ecosystem. This is of considerable importance in evaluating 'sustainability'.
Source: Gligo 1985, UNEP/CEPAL

greatest dependence on commercial energy sources, such as the production of apples, spinach and tomatoes in the United States (8, 9 and 10), show a relatively inefficient use of energy and often impose an irrecoverable burden on natural resources. The commercial viability of these crops is linked to the undervaluation of groundwater, which is treated as a 'free good'. The real cost of producing cheap spinach or tomatoes in the United States lies in depleted groundwater and increasing toxic uptake in soils and drinking water.

From the standpoint both of energy efficiency and the productivity of the agricultural system (in energy terms) per unit of land, the most successful systems are those which combine crop rotation and fallow with low energy inputs, such as the peasant maize-production system dominant in most of rural Mexico (2). This system gives an energy-efficiency factor of 10:1, fifty times that of irrigated spinach cultivation in the United States. It is worth recalling moreover that the Mexican agricultural sector is frequently thought to be languishing because of the inefficiency of peasant production, rather than of the high cost of irrigated capital-intensive agriculture which is usually invoked as part of the 'solution' to Mexico's ills (Yates 1981).

Calculations such as those in table 2.4 should cause us to question both the desirability and the inevitability of agricultural 'modernization'. First, any increased reliance on energy-intensive modern agriculture would require a concomitant increase in oil imports for less developed countries which (unlike Mexico) import most of their oil. Second, energy-intensive 'modern' agriculture uses inputs that are high-priced and beyond the reach of the vast majority of the rural population. As we shall see in chapters 4 and 5, transferring energy-intensive agricultural production to the South is favoured by larger farmers, urban interests and the multilateral suppliers of agrochemicals and machinery who often stand to benefit from the cost–price squeeze on small farmers. As Hosier et al. (1982) argue, the conventional wisdom is that the energy crisis in the North is separable from that in the South. The crisis in the North is one of oil-price increases; the 'other energy crisis' in the South is one of firewood. However, this does not mean that improved conservation of oil-based energy is equally important for both areas. As table 2.5 shows, the developed countries are much more heavily dependent on oil for energy, especially if the calculations are made on a per capita basis. It is essential for the developed countries to adopt energy-conservation policies if the consumption of a non-renewable resource is to be controlled (apart from other environmentally desirable effects that such a policy would bring). For the less developed countries one of the most potent arguments against using more oil-based energy is that, apart from its financial cost and its ecological effects, such energy feeds technological practices that make agriculture less – rather than more – energy efficient. The use to which energy is put in the process of agricultural development is only one of several dimensions of 'sustainability', but it is important that, having recognized the primary ecological meaning of the concept, we also embrace this secondary meaning and acknowledge its implications in economic and social terms.

Table 2.5 North/South oil imports

| | Oil-importing less developed countries | | Oil-importing developed countries | |
	1975	*1985*	*1975*	*1985*
Oil imports (million barrels/day)	3.12	4.35	28	35
Population (thousands)	2700	3500	590	630
Oil imports (kg per capita)	55	59	2247	2629

Source: Mackillop 1980

Population growth and carrying capacity

This chapter has discussed sustainable development largely from the standpoint of natural resources and the use that is made of these resources. The concept of 'sustainability' makes little sense, however, unless we also consider the impact of rapid population growth on the physical resource base. Clearly the potential exists for more energy-saving technologies to feed more people, and for successful replacement of the natural resources we use without accompanying environmental degradation. To what extent is the net increase in population throughout the developing countries a major obstacle to the realization of sustainable objectives? Is the demographic crisis, especially in Africa, in fact an environmental crisis from which it is impossible to recover?

Population trends for the developing world provide a bleak picture so far as 'sustaining' people at an acceptable quality of life is concerned. World population approached 4 billion in 1975; it is expected to double and reach 8 billion by 2025. According to United Nations projections the annual increase of about 80 million people will continue to grow during the 1980s until it peaks at close to 90 million, roughly equal to the present population of Bangladesh, around the year 2000 (UN 1981).

In terms of sustainability what matters most is not so much the net increase in population at the global level, but the rate of change in population in the most critical regions. At the moment population growth rates tend to be highest where basic needs are not met, particularly in Africa where per capita food production has declined by 10 per cent since 1970. Thus the current global rate of population growth of about 1.7 per cent (2.1 per cent in less developed countries and 0.6 per cent in developed countries) masks enormous regional variations (see table 2.6). Will there be a total seize-up in those countries with the most acute environmental problems well before these long-term projections materialize?

Table 2.6 Projected changes in population, 1975–2000 (millions)

	Total population 1975	2000	Net growth 1975–2000 (%)	Average annual rate of growth (%)
Developed countries	1131	1323	17	0.6
Less developed countries	2959	5028	70	2.1

Source: Global 2000 (1982), projections using medium series.

Certainly the fastest population growth rates often occur in countries with low per capita income, poor social capital, heavy reliance on labour-intensive agriculture and weak institutional infrastructures. Under the low-input farming systems that dominate most areas of Africa there is little likelihood that sustainable development can be achieved within the present context. Almost half the total land area of Africa cannot support even its 1975 population with low inputs. The critical areas include much of North Africa, almost the whole of the Sahel, the most densely populated parts of East Africa, and a dry zone stretching across southern Africa. By the year 2000 no less than 30 of the 51 countries in Africa will be unable to feed their populations with low inputs. Their total population will be 477 million, that is 58 per cent of the regional total, of which more than one half will be in excess of the land's carrying capacity.

In fact, as Norman Myers (1985) has pointed out, demographic projections make little sense against this backcloth, they 'depend on large numbers of people finding the wherewithal to sustain themselves throughout a normal lifespan' (Myers 1985, 7). These projections are *not* predictions, still less forecasts, and, although they take account of an expected decline in fertility, itself a reflection of socio-economic condition to some extent, they are not based on any consideration of the natural-resource base that sustains human societies. As Myers (1985, 7) insists, demographic projections:

> generally assume there will be some increase in human welfare, achieved in part through enhanced utilization of natural resources, and with positive implications for fertility rates. But it is beyond their scope to recognize that there can also be misuse and over-abuse of natural resources, leading to little advance in human welfare, even to declines in welfare, and with negative implication for fertility decline.

Table 2.7 Projected changes in global vegetation and land resources 1975–2000 (millions of hectares)

	1975	2000	Change	Percentage change
Deserts	792	1284	+492	+62
Closed forests	2563	2117	−446	−17
Irrigated areas	223	273	+50	+22
Irrigated area damaged by salinization and related problems*	111.5	114.6	+ 3.1	+ 3
Arable land	1477	1539	+ 62	+ 4

Note:
* Estimated as follows. It is estimated that half the world's total irrigated area is already damaged: thus the 1975 figure is approximately 111.5 million ha. The UN estimates (*Desertification: An Overview*, UN Conference on Desertification 1977, 12) that approximately 125,000 ha are degraded annually due to waterlogging, salinization and alkalinization. If it is assumed that this annual figure remains constant to the year 2000, a total of 3.1 million ha would be added to the damaged area, bringing the total to 114.6 million ha.
Source: Global 2000 (1982), projections

Projections of losses in global resources and environmental degradation (table 2.7) are as tentative as those for population and need to be
read alongside population figures. The scale of the problem can
hardly be exaggerated, however, as we will see in the next chapter.
So far as our definition of sustainability is concerned, it is essential to
consider the loss of topsoil, for example, as materially reducing sustainability. The global loss of topsoil is estimated at 22.7 billion tons a year.
This rate means that, by the end of the century, there will be one-
third less topsoil per person throughout the world than at present.

It is important to ask whether countries like India, which recently
achieved self-sufficiency in food at the national level, will be able to
maintain that achievement given that 4.7 billion tons of soil are lost
in India each year, more than in any other country. In place of 'food
security' we might argue, like Myers (1985), that 'environmental
security' is even more pressing. The illusory pursuit of 'food
security' in North America and Western Europe has helped to
produce regional structures which are a major impediment to greater
self-sufficiency in food production in the South. At the same time
the absence of 'environmental security' in the South represents an
enormous threat to the achievement of real food security in
developing countries. Nevertheless, it receives much less attention
than the threat from military aggression with which environmental
insecurity is often linked. Until we are prepared to define
sustainability in ways that take stock of both the external threat from
food policies in the North and the internal threat from demographic
pressure in the South, it will remain something of a chimera. It
remains to consider to what extent the discussion of sustainable
development has incorporated an awareness of these issues.

Sustainable development: a new paradigm?

The term 'sustainable development' was used at the time of the
Cocoyoc declaration on environment and development in the early
1970s. Since then it has become the trademark of international
organizations dedicated to achieving environmentally benign or
beneficial development. The term has served to catalyse debate over
the relationship between economic change and the natural-resource
base in which it is grounded, especially in the publications of the
International Institute for Environment and Development (IIED).
The term 'sustainable development' suggests that the lessons of

ecology can, and should, be applied to economic processes. It encompasses the ideas in the World Conservation Strategy, providing an environmental rationale through which the claims of development to improve the quality of (all) life can be challenged and tested. To what extent though does sustainable development provide an alternative paradigm, or system of meaning, as well as a focus for improving environmental policy and management?

In describing what he calls 'co-evolutionary' development, Norgaard (1984a and b) outlines possible linkages between economic and ecological paradigms. His intention is not to construct a new paradigm out of what remains of existing models, but to draw attention to the advantages of using different models simultaneously:

> A linkage is quite different from a grand synthesis of previously incongruous paradigms. Through a linkage, each discipline enriches the other because of their differences. Neither discipline must abandon its past. Eventually, however, new emphases and approaches arise because of the enrichment . . . (1984a, 525)

Norgaard calls his co-evolutionary approach a perspective, rather than a paradigm. As we shall see in chapter 3, this perspective takes issue with the way in which economics tackles environmental processes, attributing the short-sightedness of many economists to their attachment to a mechanistic view of science. The implications of Norgaard's perspective are much more radical than the injunction to consider different disciplinary perspectives originally suggests.

Sustainable development requires a broader view of both economics and ecology than most practitioners in either discipline are prepared to admit, together with a political commitment to ensure that development is 'sustainable'. The practical implications of such a position are important and cannot easily be avoided. Is it possible to undertake environmental planning and management in a way that does minimum damage to ecological processes without putting a brake on human aspirations for economic and social improvement? Does sustainable development have a methodology and a praxis? These questions await discussion in later chapters. For the moment it remains to discuss the ways in which the sustainable-development approach has itself been revised and refurbished to take greater account of the underlying inequalities that limit the livelihood opportunities of poor people and their environments.

Ecodevelopment

One approach that has gained some currency is that of 'eco-development'. The advocates of ecodevelopment include those who regard an alternative approach as essentially political. The objective is not merely to identify the limitations of existing approaches; it is to advocate alternatives that deal effectively with 'the power variable':

> Naive statements on needs, participation and environmental compatibility are espoused in many papers . . . But whose needs are going to be met and whose are not; who will participate and who will not; and which lobbies, interest groups, and economic and political entities will be hurt by environmental compatibility? (Farvar and Glaeser 1979, 1)

Taking a long look at the practical commitment of international organizations to sustainable-development objectives Farvar and Glaeser conclude that, even when fundamental approaches like land reform and restructuring of the relations of production have been considered in the agenda of such organizations, 'the real issues have been obscured and neutralized by sterile language and wrong premises' (Farvar and Glaeser 1979, 6). They attribute this ineffectiveness to a number of factors, among them budgetary cuts (for example, the budget of the United Nations Environment Programme (UNEP) was halved between 1975 and 1979), and to the antipathy of the various 'super powers' to any programme based on self-reliance and the reduction of the South's technological dependence on the North. Clearly advocating sustainable development or ecodevelopment in principle does not commit governments or international organizations to its achievement in practice.

'Ecodevelopment' is also the term given to the planning concept originally advocated by UNEP. It was defined as 'Development at regional and local levels . . . consistent with the potentials of the area involved, with attention given to the adequate and rational use of the natural resources, and to applications of technological styles' . . . (UNEP 1975). This usage suggests a regional focus for resource planning, informed by technological considerations. It is, as Bartelmus (1986) observes, a long way from the ethically committed, integrated approach suggested by Riddell (1981). It is also a much more limited concept of 'ecodevelopment' than that proposed by

Sachs (1984), for whom international structures, as well as moral commitment, need to be radically changed. It was a non-governmental organization, the Centre International de Recherche sur l'Environnement et le Developpement (CIRED) which pioneered pilot studies in developing countries, in which alternative energy sources, recycling and organic agriculture played a considerable part (Sachs 1976, 1980).

The practical attention which CIRED gave to environmental alternatives was matched by a concern to incorporate social and cultural processes within the ecodevelopment approach. In its early stages this had the rather benevolently paternalistic imprint of so much planning methodology: 'preparatory education to create social awareness of ecological values in development . . . resource development for the satisfaction of basic needs . . . the development of a satisfactory social ecosystem'. Nevertheless, the necessity to incorporate social variables has remained with the advocates of ecodevelopment. Dasmann, for example, departs from UNEP's planning approach in advocating little short of moral renewal and self-sufficiency: 'To somewhat simplify ecodevelopment, I have considered it to be represented by a triangle, one side of which is *basic needs*, the second *self-reliance*, and the base *ecological sustainability*' (Dasmann 1985, 215). Each of these variables needs to be considered, since development will not be sustainable unless poor people are involved in meeting their aspirations. As Chambers has expressed it, what is required is that 'last' thinking is adopted, putting people first and poor people and their priorities first of all. 'The environment and development are means, not ends in themselves. The environment and development are for people, not people for environment and development' (1986, 7). In Chambers' terminology sustainable development is a 'first' concept, since it has emerged from the offices of IIED and the discussions of the Brundtland Commission (WCED 1985). In his view the poor are largely concerned with their immediate livelihoods; it is the enlightened rich who give priority to sustainability. What is required is a mental leap along the lines of 'sustainable livelihood thinking' (Chambers 1986, 10). He contrasts 'environment thinking' with 'development thinking' and 'livelihood thinking'. The perspective of the poor is at variance with that of most economists and biologists, placing the immediate satisfaction of needs and the avoidance of risk before sustainability or higher productivity. Similarly the time horizon of the poor is shorter, the future valued much less than the present. According to Chambers, 'sustainable livelihood thinking' enables causal connections to be made between development and livelihoods and between the environment and livelihoods. What poor people pursue through the development process and their use of the environment, is simply a better livelihood.

The perspective advocated by Chambers is a real advance on much of the 'institutional' writing about sustainable development. As we have seen, the concept has both economic and ecological parameters which are difficult to marry. Dissatisfaction with the narrowness of the ecological model has led to attempts to steal or borrow some sociological content, either through social planning norms (UNEP's use) or through referring to 'basic needs' and 'self reliance' (Dasmann 1985). Chambers, rightly in my view, criticizes the suggestion that we are in a better position than the poor to recognize what is good for them, an assumption contained in much of the 'basic needs' writing. He correctly asserts that 'short-term improvements in living create conditions for later livelihood-intensive human use of the environment which is sustainable' (1986, 13). Human welfare should be the point of departure.

The nagging doubt remains however; despite the seriousness of Chambers' position, it does not represent an adequate response to the issue raised by Farvar and Glaeser (1979). For sustainable development to become a reality it is necessary for the livelihoods of the poor to be given priority, but how can this priority be pursued at the local level while the effects of international development systematically 'marginalizes' them? The political aspects of development extend to sustainable development options which can only be achieved through political changes at the local, national and international level. Chambers believes that 'political economy' can be incorporated in his approach by examining the net effect of transnational corporations and logging contractors on sustainable livelihoods (1986, 13). However, these interests are often determining factors in the availability, or otherwise, of sustainable livelihoods. It is just as necessary to put political economy first as the real-world thinking and priorities of the poor. As we shall see in chapter 4, unless we pitch our conception of sustainable development at a level which recognizes international structures, it is in danger of being yet another discarded development concept. Its polemical usefulness will have outlived its practical utility.

3

Economic models and environmental values

Before examining international restructuring and its local consequences, in more detail, it is important to address the theoretical
concern that lies at the heart of the economic–development process.
To what extent has economics been able to incorporate environmental considerations within its governing paradigm? In drawing
attention to ecological factors and their relationship to social
structure, we need to be able to specify the circumstances under
which they are taken into account and become part of the economic
modelling process. Is it possible to 'internalize' the externalities that
economists identify with the achievement of economic growth, and
that, as we have seen, seriously handicap the capacity of poor
households to reach sustainable objectives? In this chapter several
different approaches to economic behaviour and environmental
values are examined: differences within neo-classical economics,
'deep' ecological positions and Marxist theory.

Environmental economics

There are several schools of thought on the relationship between the
environment and economic growth. They range from what O'Riordan
(1986) has termed the 'environmental moralists' who deny that the
environment is a commodity at all to those who argue that
environmental goods should be treated exactly like any other
commodity for which there is a market. In the view of some writers,
neo-classical economics has 'largely been devoted to the refinement,
expansion and implications of thinking of the environment as a
commodity' (Pearce 1985, 9–10). Other economists, such as
Norgaard, point to the difference between examining *how* scarce
resources can best be allocated (the definition of most economists

today) and turning this framework on its tail 'to determine from how resources are allocated *whether* they are scarce' (Norgaard 1985b, 3). Economists, in other words, are interested in scarcity as the underlying reality behind human choice. Environmentalists are concerned that economic growth is the reality which makes human choice less and less possible under conditions of scarcity.

Economists like Pearce argue that it is possible to consider the environment within the governing economic paradigm, and that the field of 'bioeconomics' has already made substantial progress. They argue that extended cost–benefit analysis is already of considerable use as a decision-making tool. Being able to quantify and measure human concern for the environment is then of major assistance to the environmental lobby, and a step forward in building bridges with non-economists interested in the environment. Pearce is concerned that the environmental movement is either oblivious of this fact or actually opposed to it, believing (wrongly in his view) that the esteem in which economics is held by planners and policy makers partly accounts for the problems encountered in the environment. The alternative view is that a concern with the environment entails the abandonment of a unitary economic paradigm. This is essentially the position of economists like Norgaard (1985a and b).

Pearce's position, like that of most of the economists he cites, is that economic modelling is increasingly sophisticated and able to attach quantitative weight to human desires and preferences. Experimental economics has led to 'the extension and measurement of our value concepts beyond those of direct use value', including critically those pertaining to environmental uses (Pearce 1985, 12). By being able to incorporate measurable data into policy work on the environment, the interests of sustainability are served, though this is not a guarantee that they will be adhered to by policy-makers. Economics is traditionally interested in value that can be expressed through consumer preferences, for example, in the notion of 'opportunity cost'. These methods can be extended to cover environmental goods in a number of ways.

There is then a recognition in this approach that any costs of development which lie outside ecological constraints, that pose difficulties for sustainability, are both injurious to life-support systems and serve to reduce the future resource base. At one level this is little more than the admission that market mechanisms and public choices do not necessarily bring about sustainable development. It is also an affirmation of the strength of economics in being able to measure the effect of these allocative mechanisms in areas, such as environmental preference, which have received attention only recently. Pearce reiterates that the pursuit of better living standards is captured in the concept of economic growth, and that it would be unwise to fail to acknowledge this fact (1985, 16).

The ability to model human preferences for environmental goods, through the 'willingness to pay' principle, rests on ways of discounting present and future preferences. The rules for assessing anticipated future losses 'can be modelled and analysed in a framework which incorporates both economic and ecological considerations' (1985, 20). The claim is that 'economic science has been ahead of the game, developing the foundations of bioeconomics, the integrative analysis of biological and economic systems of man (sic) and the natural environment' (1985, 25).

Pearce's position can be criticized from within economics, as well as from without. A recent paper concerned with developing methods of environmental evaluation on the basis of people's choices, concluded that there were important divisions within economics '. . . not only as to the appropriate rate of discount but even as to whether long term environmental changes should be the subject of discounting at all' (Hodge 1986, 9). Many economists remain unconvinced that environmental resources represent a challenge which neo-classical economics can and should address. Some, following a utilitarian position, would argue that there is no moral justification for extending individual rights to future generations, for example. Some economists clearly do not see the environment as a problem for economics, even if economics is a problem for the environment.

Another position, and one which takes issue with much that Pearce is saying, argues that economists are still giving the environment much less attention than it deserves. In a series of trenchant, closely argued papers Norgaard (1984a, 1985a and b) has suggested that neo-classical economics is incapable of fully incorporating environmental considerations into its methodology without what amounts to a 'paradigm shift' (Kuhn 1962). Norgaard insists that economic models do need to meet the challenge of future discounting. For one thing, future generations need to inherit an improved capital stock and better technology that will equip them to substitute resources and overcome scarcity. The need to treat future generations as if they are living now, he argues, is not just a requirement of equity, but of the competitive conditions assumed by the economist's model which assumes exchange between generations. He notes that the world which has been formally modelled by economists is an '. . . imaginary world without surprise or resources'. It is also a world which assumes general public policies and allocations by public agencies which few of the economists concerned would be prepared to defend for political reasons (1985b, 4).

The problem with environmental resources is partly that of determining optimal behaviour. Allocations which account for environmental preferences, in the way referred to by Pearce, frequently assume that behaviour is optimal when it is not. For example, people have inadequate knowledge of the decision-makers whose behaviour is altering their environment, and little knowledge of whether their behaviour is appropriate. Since many natural resources are located on public lands, the assumptions of the economic model also frequently concern the public's knowledge of how public agencies work. This leads us into particularly dangerous waters. If, on the other hand, economic indicators give the correct signals without considering new areas of behaviour and new preferences '. . . then resource owners are already fully informed about scarcity' and are practising 'optimizers' (Norgaard 1985b, 6). Norgaard argues, convincingly I believe, for a much more ecumenical view of the environment and development in which long-run resource scarcity is considered together with the technological drive to offset it. In his view developing and adapting to technology 'have restructured what is a benefit and what a cost' in the environment (1985b, 14).

The problem of marrying economics to the environment is not confined to methodological postures, however important these may be. It begins with the assumptions of the disciplines concerned. Neo-classical economics assumes that resources are divisible and can be owned. It does not acknowledge that resources bear a relationship to each other in the natural environment, as part of environmental systems. Market mechanisms fail to allocate environmental goods and services efficiently precisely because environmental systems are not divisible, frequently do not reach equilibrium positions and incur changes which are not reversible. In other words, the properties of ecological systems run counter to those of what Norgaard terms 'the atomistic-mechanical world view' of neo-classical economics (1985b). Economics is not adapted to consider total changes. Resting as it does on the concept of the margin, it is epistemologically predisposed towards a reductionist view of resources and their utility.

Similarly economic theory had difficulty in recognizing that both ecological and social systems evolve over time, in ways which change both of them. This evolution brings uncertainty and the uncertainty of evolving systems is not adequately accounted for by economists interested in risk within a neo-classical model. The implication is that economics can only handle environmental factors successfully if it breaks free from its mainstream epistemology.

This is not to argue that human environmental preferences cannot be modelled by economists, or anyone else. Such modelling, however, is only useful when 'environmental' goods can be clearly distinguished from other goods. This is frequently not the case. Moreover, human preferences for environmental goods need to take account not only of the value of the environment to human beings, but also the value of the environment itself. Ecosystems are themselves a source of value. The 'deep ecology' position addresses just this concern, as we shall see below. Finally, the real world of resources and the environment only weakly resembles that of the economists' models, since the environment is constantly evolving together with the social system in ways that alter its nature, whether human actors are aware of it or not. In discussing genetic engineering, for example, in chapter 8, we will be examining an area in which environmental changes are occurring without the knowledge of anybody but a specialized, scientific élite. For the moment, however, we need to examine the question of environmental values in more detail.

Environmental goods and values

The discussion of growth takes on a completely different character if we do not regard the market as the ultimate barometer of peoples' needs. Writers like Maslow (1954) have argued that 'needs' can be ranked in different societies. The priority of peoples' needs changes in the course of development, from the satisfaction of basic needs such as food and shelter, to the satisfaction of aesthetic and existential needs (or wants) which play such a large part in developed countries. In poor countries environmental goods are 'survival goods': fuelwood, clean water, staple food supplies. In developed countries what Hirsch (1976) calls 'positional goods' play an increasingly important part in personal well-being. The countryside is a good example of a 'positional good' in that its value declines as access to it increases. In a sense, then, it is subject to market mechanisms but differs from most commodities (although not all) in that it cannot be produced in larger quantities to satisfy demand. The challenge of environmental social science is to link the priority of needs to their conditions of scarcity (Inglehart 1981).

We need to know to what extent economic growth makes an increasingly marginal contribution to peoples' needs but at the same time creates scarcity where it did not exist before. The observation that an inverse relationship exists between human preferences for goods and the likelihood that they will become more scarce is not confined to environmental goods, but the environment is an area in which these considerations loom large. This is also an area in which social scientists, including economists, have a large role to play (Lowe and Rudig 1984).

One of the things which distinguishes the late 1980s from the 1960s is that today, in most parts of the developed world, the pursuit of economic growth and the production of more goods is being effected through cuts in the contribution of the public sector, in relative if not in absolute terms. Public-sector cuts have reduced social-service provision, in many cases bringing a decline in the slow process of inner-city rehabilitation, in industry in the regions and, most dramatically, in employment. Public opinion in the industrialized countries is nevertheless favourable to environmental protection – placing it before giving priority to economic growth (see table 3.1). Less developed countries, as we shall see in chapter 4, have been affected by the same processes, often in an exaggerated form as passed on to them by international financial institutions and the increasing protectionism of the industrialized countries. The technologies which are credited with the potential for reversing the depressed pattern of economic growth, microelectronics and biotechnology, for example, are unlikely to reverse the environmental effects which have accompanied a restructuring of the economies of the developed countries. Indeed the expansion of new industries in most industrialized countries is being achieved by depressing public expenditure still further in areas of environmental concern. One regional example of these restructuring trends on the environment in

Table 3.1 Public opinion: environment protection vs. growth trade-off, selected industrialized countries

	Year	No. of inter- views	Priority to environ- mental pro- tection (%)	Both are possible (%)	Priority to economic growth (%)	Don't know (%)	Total (%)
USA	1984	1590	62	n.a.	28	10	100
Japan	1981	2426	28	41	11	20	100
Finland	1983	2050	47	35	11	7	100
EEC							
Belgium	1982	1020	50	n.a.	30	20	100
Denmark	1982	995	75	n.a.	14	11	100
France	1982	939	58	n.a.	30	12	100
Germany	1982	1012	64	n.a.	21	15	100
Greece	1982	1000	56	n.a.	26	18	100
Ireland	1982	1007	29	n.a.	58	13	100
Italy	1982	1025	67	n.a.	20	13	100
Luxembourg	1982	300	64	n.a.	26	10	100
Netherlands	1982	1056	56	n.a.	34	10	100
UK	1982	1335	50	n.a.	36	14	100
EEC total	1982	9689	59	n.a.	27	14	100

Notes:
Results from public opinion surveys, carried out in the United States, Japan, Finland and EEC countries, within a co-ordinated framework suggested by OECD, are presented here.
n.a. = not asked.
The table shows public-opinion data concerning environmental protection versus economic growth trade-off, according to four response categories in percentages of the total number of persons interviewed.

less developed countries is that of Latin America, discussed later in this chapter.

An example of an environmental philosophy with an epistemology radically different from that of 'bioeconomics' is that of 'Deep Ecology'. Deep Ecology is the name given to the philosophical position of a series of writers whose roots lie in Scandinavia, California and Australia (Naess 1973 and 1983, Tobias 1984, Sylvan 1985a and b). According to its adherents, Deep Ecology is metaphysical at base, it represents a search for a sustaining metaphysics of the environment. The underlying conviction that informs this view is that human beings should seek to emphasize their underlying unity with other living beings and processes. Deep

Ecology is biocentric, not anthropocentric (Sylvan 1985a, 2). Unlike reformist environmentalism, Deep Ecology is 'not a pragmatic, short-term social movement with a goal like stopping nuclear power or cleaning up the waterways' (Devall 1979). The defining characteristic of 'deep', as opposed to 'shallow', environmental positions is that they do not take an 'unrestricted' view of the purposes to which the natural environment can be put. This is in contrast with dominant Western culture which emphasizes that people can do more or less what they like with nature, which exists for humans to exploit or manage. According to 'deeper' positions, humans are not the sole items of value which bestow value in the world, 'and not all things of value are valuable because they answer back in some way to human concerns' (Sylvan 1985a, 5).

Deep Ecology has a small following, partly because some of its tenets, such as biospheric egalitarianism, can easily be dismissed as eccentric or untenable. It is not obvious, as Sylvan points out, that something has value to humans simply because it has life. Nevertheless elements of a deep ecological concern can be identified in a much wider range of writing and thinking about the environment (Cotgrove 1982, Russell 1982, Rifkin 1985). The existence of philosophical positions like that of Deep Ecology and of social movements committed to 'deep' ecological objectives like the animal-rights movement should serve to remind us of the impossi-

Table 3.2 Typical components of growth/environment paradigms

Dominant social paradigm	*Deep ecology paradigm*
Dominance over nature	Harmony with nature
Natural environment as a resource	Values in nature/biosphere impartiality
Material goals/economic growth	Non-material goals/ecological sustainability
Ample reserves/perfect substitutes	Finite natural reserves
High-technology/science solutions	Appropriate-technology solutions
Consumerism	Basic needs/recycling
Centralised/large scale	Decentralised/small scale
Authoritarian/coercive structures	Participatory/democratic structures
Shallow	*Deep*

Source: adapted from Sylvan 1985b (2), 12

bility of incorporating environmental values fully within the economic paradigm.

Table 3.2 sets out some of the contrasts between the 'deeper' ecological paradigm and the dominant social paradigm which, among other things, places considerable emphasis on economic growth. It should not be taken to imply that those who make use of neo-classical models are necessarily opposed to the social values which are located on the deep-ecology side of the diagram. Most conservation positions occupy the shallow ground between this dominant social paradigm and Deep Ecology. Setting out the variables in this way enables us to bring together the different components of these paradigms, and serves to illustrate the multifaceted character of environmental value systems.

Marxist perspectives

Economic development is not the province of neo-classical economics alone, any more than a concern for the environment is the sole province of environmentalist positions. Contending theories of economic development necessarily include Marxist perspectives. Marxist analysis has traditionally looked upon environmental problems as a necessary, but unfortunate, consequence of the development of capitalism. This position has proved unsatisfactory for a number of reasons. First, it is clear that such problems are not confined to capitalist societies, despite the elaborate and disingenuous claim that Soviet society takes ecological issues seriously (Khozin 1979). Second, it has become increasingly clear that '. . . the goal of expanding the productive forces is in conflict with the original revolutionary goals of eliminating exploitation and alienation' which are central to Marxism (Lashof 1986, 13). Since Enzensberger's seminal essay on the subject, a new generation of the European Left has claimed to be both socialist and pro-ecology (Enzensberger 1974). Third, it has become increasingly clear that the ecological breakdown, forecast in the 1960s and 1970s, has already occurred in parts of the South, providing a curious footnote to Marx's stark choice between socialism and barbarism. Finally, and most importantly, Marxist theory and method, divorced from orthodox dogma, still represents one of the most fertile intellectual traditions in which to locate ecological ideas, based as it is upon both the social construction of nature and the 'naturalization' of human consciousness (Schmidt 1971, Smith 1984).

A Marxist view of the environment needs to encompass a number of closely related, but separable issues. In the first place it needs to address the issue of the way nature is transformed under capitalism and the implication of this process for developing countries today. In this context the question of commodity production is of paramount importance and distinguishes Marxism from both neo-classical and radical ecology perspectives, as we shall see. Second, Marxist approaches to the environment are necessarily concerned with the distribution of environmental costs and benefits, not simply from a welfare standpoint, but because the distributive effects of environmental change have important implications for the kinds of social movements which are likely to emerge from ecological degradation. Third, Marxists are concerned about the ideological content of environmental ideas and their relationship both to bourgeois processes of legitimation and to central Marxist concepts such as alienation and the class struggle.

The concept of 'the environment' appears in Marxist writing in a number of guises. The most common is that of 'natural resources'. As Schmidt (1971) has argued, natural resources are the product of a conversion process through which labour (and capital) is applied to nature. There is nothing 'natural' about natural resources to begin with – this property is socially determined in any given environment. Natural resources are those which are of potential use to human beings. They are socially determined in the sense that their value is related to the technologies used to exploit them and the existence of people to consume them. Without a social formation comprising, among other things, an internal market of consumers and a technology capable of linking them to the production process, natural resources remain 'unresourced'.

This leads to the question of underdevelopment. As Ojeda and Sanchez (1985, 36) have put it:

> private capitalist accumulation, within the ambit of the international division of labour and the growing specialisation that this implies, makes the consideration of what is a natural resource depend upon the manner in which each society is inserted within the world market . . .

Capitalist development transforms nature and the environment within a logic which needs to be understood in global terms, as both

Lenin (1972) and Luxemburg (1951) argued, and which has characteristics today which it did not possess 50 or 100 years ago. This internationalization of the environment within the global capitalist system is examined in chapter 5. The implication of viewing resource uses as the conservation of *stocks* rather than the utilization of *flows* has informed some important work currently being undertaken in Latin America by the United Nations Economic Commission for Latin America (ECLA) and the United Nations Environment Programme (UNEP) (Sunkel and Gligo 1980), which is discussed in chapter 5.

To some extent Marxists encounter the same problem in approaching the environment as neo-classical economists. This is that the environment exists as a system. It follows that '. . . a basic difficulty in the construction – or refutation – of ecological hypotheses is that the processes involved do not take place serially, but in close interdependence' (Enzensberger 1974, 6). As we saw, the systems in which resources are located provide difficulties for neo-classical economists whose paradigm is atomistic, if not reductionist. For Marxists the problem is one of the underlying logic of the environmental system rather than the system itself. Neo-classical economists have attempted to get around the problem of environmental values by attaching a price to 'externalities', enabling them to be treated as if they were part of an optimizing resource model. This does not necessarily enable economists to incorporate the environment successfully within their analysis as we have seen, but it does enable them to model human preferences in the environment, at least at a theoretical level.

The problem for Marxists is more complex, however. As Harris has written, Marxist methodology '. . . points in an opposite direction. It suggests that it is not possible to flout the market system. If a system of commodity production prevails, merely raising the price of some commodities will not eliminate them from use . . . It will not change people's relation to nature, their attitudes or their desire for material possession' (Harris 1983, 49). In Marxist terms scarcity will disappear only when the necessity to make commodities in order to realize a profit disappears. While some neo-classical economists assert that environmental goods should be approached for their use values as well as their exchange values, Marxists are concerned, primarily, with the process through which use values are converted to exchange values. The point of Marxist analysis is not to espouse economic growth as an end in itself, but to argue that the increased production of commodities under capitalism necessarily implies economic growth. Together with radical ecologists, Marxists agree that the market allocates natural resources in an inefficient way through time, ultimately destroying the basis of survival for future generations. Nevertheless, Marxists see the commitment to commodity production under capitalism as making ecological externalities inevitable. Indeed it is part of the contradictory nature of capitalism that 'the environmental crisis presents a massive threat' to the earning powers of entrepreneurs, as underwritten by the capitalist state (Enzensberger 1974, 11).

The inevitability of ecological collapse was not a concern of early Marxist writing, although Marx referred to the problem of maintaining soil fertility in a celebrated passage from *Capital* (1974). Similarly Engels (1970a and b), although aware of the fact that nature could not be plundered without cost, still emphasized the potential that nature afforded economic growth. What distinguishes the position adopted by radical ecologists, and some Marxists from that of orthodox Marxism, is the conviction that '. . . long before it becomes physically impossible to grow, it becomes socially undesirable to do so' (Lashof 1986, 10). The costs of environmental degradation, especially in the South, are such that the final scenario of capitalism destroying itself through ecological attrition is unacceptable in this analysis. The point at which the costs in destroying the environment and non-market social relationships exceed the benefits of further commodity production has already arrived. The Promethean quality of early Marxism is therefore placed in doubt. As Enzensberger wrote during the 1970s (1974, 3), when radical ecology was primarily a critique of 'successful' economic growth:

> . . . (the ecologists) have one advantage over the Utopian thinking of the Left in the West, namely the realization that any possible future belongs to the realm of necessity not that of freedom, and that every political theory and practice – including that of socialists – is confronted not with the problem of abundance, but with that of survival.

These sentiments showed some prescience in 1974 when they were written. After the recurring crises of African famine, and the nuclear catastrophe which occurred at Chernobyl in the Soviet Union, they look increasingly realistic. Clearly environmental disasters can anticipate the point at which the limits to economic growth are finally reached.

In this context it is worth reflecting that the production of commodities under capitalism can be taken to include 'free goods' in nature which originally had no market value. The transformation of nature under capitalism has largely been conceptualized in terms of the specialization of labour and technology. Human dependence on technology has already reached the point, however, where nature itself can be maintained only through recourse to technology. For example, chemical fertilizers and pesticides are necessary to agricultural production under some conditions, whatever their long-term effect. Similarly, even air and water cease to be 'free goods' in a pure form and become highly valued commodities when the environment is sufficiently degraded.

This brings us to consider distributional aspects of the environment which are central to a Marxist approach. During the nineteenth century most ecological degradation was more class specific in the industrialized countries than it is today. The housing and working conditions of the poor were extremely bad and lay outside the experience of the middle and upper classes. The circumstances in much of the South today are comparable, in that environmental problems are differentially distributed in ways that can hardly be overlooked. One view is that, as environmental problems in the North have tended to affect the affluent as well as the poor, ecological issues have caught the attention of the middle classes. The 'cost of a private environment' is 'already astronomical' in the industrialized countries (Enzensberger 1974, 10). The environmental movement, it is argued, has great appeal to those who can do little to remove themselves from environmental hazards such as acid rain, nuclear fallout or industrial pollution. The Club of Rome, it should be remembered, was composed of top industrialists and urban bureaucrats who were also concerned at the cost to the quality of *their* life, posed by pollution, industrial waste and urban decay. Today this list of the undesirable consequences of economic growth would have to be lengthened to include some additional factors: the survival of natural species, the difficulty in gaining access to the countryside and the erosion of farmlands in parts of North America and Western Europe.

The importance of distributive factors in the way environmental processes are undermined has been the focus of considerable attention (Sandbach 1980, Sandbrook 1982) but it requires a much more systematic analysis than Marxists have hitherto provided. Enzensberger, for example, argues *both* that the environmental crisis is a threat to the earning power of the bourgeois (1974, 11) *and* that 'the eco-industrial complex' is able to profit from pollution at the expense of the community as a whole (1974, 12). These observations, which appear contradictory, require careful separating. In the late 1980s the political assumption is that the earning power of the capital-owning classes is being put at risk by 'expensive' controls to protect the environment, rather than by degraded environments.

Similarly Enzensberger's observation that bourgeois social movements became interested in the environment only when the middle classes were adversely affected conceals as much as it reveals. One might equally argue that Marxists were not interested in the environment until it became an issue in bourgeois political debate. The environmental problems which were the product of class oppression assumed political importance for many Marxists only because the movements which sought to improve environmental conditions since 1945 evolved largely independently of the labour movement. At the same time radical environmental groups, such as the Green parties in West Germany and the Netherlands, have made the distributive consequences of economic growth for the environment a central part of their analysis and their political manifesto.

The full impact of Marxist thinking on the relationship between society and nature will inevitably depend on the degree to which Marxism fully incorporates the implications of unsustainable development as we shall see in chapter 8. It will be clear that the concepts which have been employed within Marxist theory to examine the role of commodity production in transforming the environment need to be elaborated in terms of the system's ability to reproduce itself, as well as in terms of production.

Capitalism necessarily concerns the transformation of nature and the development of value around market conditions of production and exchange, but this has become divorced from what Burgess (1978) identifies as the neo-idealist concern with human consciousness. The Marxist concept of nature should aspire to make the connections 'between attitudes towards nature and the day-to-day practice of man's [sic] economic activity', as Pepper (1984, 159) argues, but this aspiration needs to be grounded in a theoretical framework. This theoretical framework, in turn, can succeed only if Marxism is prepared to question some of the assumptions of nineteenth-century theory, while making full use of its method. Enzensberger (1974) has criticized the ecological movement for its lack of ideological sophistication, declaring that it is ill-equipped to make the transfer from the natural world to its social mediation. Radical ecology, he argued, has no theory of society and no sense of the historical process.

However, the central contradiction of advanced capitalism and its relations with the developing world still eludes Marxism. This is that the labour process, the means by which the social mediation of nature is achieved, succeeds in transforming the environment in ways that ultimately make it less productive. At the same time the 'externality' effects that have attracted the attention of economists in western societies are far more central to the 'survival algorithm' of many households in the South than many Marxists have acknowledged. Technology is not simply a means to harness nature in industrial society, it is also the instrument through which people can become alienated from nature in rural areas of the South. People are frequently separated from their land; women from their control over household resources; cultural practices that have evolved to sustain both production and the environment are lost. In order to appreciate fully how this occurs we need to turn our attention to the international system and the historical processes which have contributed to the environmental problems we recognize today.

4

Sustainable development: the problem

We have seen that sustainable development can be approached from a number of directions. The advocacy of sustainable development in documents such as the World Conservation Strategy, however, fails to come to grips with the central issue of economic growth, as the motor behind development. The discussion of sustainable development was principally addressed to the negative consequences of development; this might meet economic criteria but seriously underestimates ecological (and social) factors. In the previous chapter we asked whether economic growth within ecological limits fully answers the call for a more resource-sustainable development. In this chapter we turn our attention to the problem of economic growth and the environment in existing societies of the South. The analysis is directed towards identifying those features of the international economy – trade, aid policy, the debt crisis and the behaviour of multinational corporations – which carry negative implications for the long-term sustainability of the development process.

The limits to economic growth

In July 1970 an international research team at the Massachusetts Institute of Technology (MIT) began a study of the effects and limits of continued worldwide growth. The Report of the Club of Rome (titled *The Limits to Growth* and published in 1972) marked the high-water mark of anti-growth sentiment. In the light of subsequent international reports, such as Brandt (1980) which highlighted the increasing disparities between North and South, and the increasingly critical message of global conservation studies (World Conservation Strategy (1980), Global 2000 (1982)), it is worth reminding ourselves of the conclusions contained in the original *Limits to Growth* document. These were:

1 If the present growth trends in world population, industrialization, pollution, food production and resource depletion continue unchanged, the limits to growth on this planet will be reached sometime within the next one hundred years. The most probable result will be a rather sudden and uncontrollable decline in both population and industrial capacity.

2 It is possible to alter these growth trends and to establish a condition of ecological and economic stability that is sustainable far into the future. The state of global equilibrium could be designed so that the basic material needs of each person on earth are satisfied and each person has an equal opportunity to realize his individual human potential.

3 If the world's people decide to strive for this second outcome rather than the first, the sooner they begin working to attain it, the greater will be their chances of success.

(Meadows *et al*. 1972, 24)

The Club of Rome's Report recommended a conscious move from exponential growth to global equilibrium. Exponential growth was dangerous partly because exponential increases were deceptive: growth approached a fixed limit (such as natural-resource endowments) with a suddenness that could easily be overlooked (Meadows *et al*. 1972, 29). It was also a dynamic phenomenon, making analysis 'of the causes of growth and the future behaviour of the system . . . very difficult indeed' (1972, 30). The world's resources are finite, and the trade-offs between human activities – the production and consumption of food, the control of pollution – met absolute limits. The *Limits to Growth* made it clear that it was not possible to foretell exactly which limits would occur first, or what the consequences would be, because of the unpredictable nature of human responses to the crisis (1972, 87). Fifteen years after the Club of Rome's Report was published it is worth asking whether anything has altered the validity of the original conclusions which, it should be remembered, attracted considerable criticism at the time (Gribben 1979).

The final communiqué of the 1985 Bonn Summit of the industrialized (Organization for Economic Co-operation and

Development countries (OECD)) countries reiterated the central beliefs of the industrial (OECD) nations in economic growth as a major ingredient in development. [The 1986 Summit of OECD in Tokyo differed little in most respects from that of Bonn a year earlier, except in recommending measures to stabilize exchange rates, a proposal that reflected a change in the United States administration's financial managers.] The Bonn Summit recommended:

1 A sustained level of economic growth in the developed world, at least 3 to 4 per cent per annum.
2 The significant expansion of world trade, enabling LDCs to increase their export earnings.
3 Moves towards open markets and an end to protectionism (by 1986 these injunctions were largely aimed at Japan).
4 The need for lower interest rates.
5 Flexible rescheduling of international debts, while LDCs began to achieve trade surpluses.

The Club of Rome had recommended an end to exponential growth as a means of recovering the equilibrium between population and resources. The objective was not to place the 'environment' above the living standards of the poor in the South, but a commitment to meet the basic needs of the poor as the prime objective of a much more limited growth trajectory. In the succeeding decade and a half the reverse has happened: international trade has become distorted in ways that have led the poor actively to 'feed' the rich. From 1981 to 1984 profit remittances from LDCs averaged US$14 billion annually. Annual interest payments on foreign debts increased tenfold, from US$2.5 billion in 1970 to $25.7 billion in 1979. They doubled again to US$51 billion in 1981. At the same time in some parts of the world food-consumption levels actually declined: per capita food production has declined in Africa by 10 per cent since 1970, the shortfall being made up by increased imports.

Since 1979 the world economy has grown at an average rate of 1.7 per cent per annum, exactly the same as population growth. This suggests that, on average, there has been no increase in world output per person (Brown 1984, 17). The factor which has slowed economic growth since the Club of Rome's Report has been the rise in the price of oil from US$2 a barrel in 1972 to an average of US$12

Table 4.1 World economic and population growth at three oil-price levels, 1950–83

Period	Oil price per barrel ($)	Annual growth		
		Gross world product	Population (%)	Gross world product per person
1950–73	2	5.0	1.9	3.1
1973–9	12	3.5	1.8	1.7
1979–83	31	1.7	1.7	0.0

Note:
Oil-price data are from International Monetary Fund; gross world product data are from Herbert R. Block (1981) *The Planetary Product in 1980* Washington, DC, US Department of State, and Worldwatch Institute estimates; population data are from United Nations and Worldwatch Institute estimates.
Source: Brown 1984

between 1973 and 1979, and US$31 between 1979 and 1983 (see table 4.1). This had led to a zero-growth rate per capita by 1979/83.

The slight contraction in world trade in the 1980s and the achievement of zero economic growth per capita have been very mixed blessings. On the one hand the fall in oil consumption, especially that of oil-importing countries, has led to energy consumption that is more efficient and, increasingly, more localized energy production. Nevertheless, as we have seen, the global imbalances remain and, in some regions such as Africa, have brought a deterioration in the living standards of the poorest. The solution to these problems will not be found in further oil-price decline. Other factors need to come into play: industrial growth needs to be redirected towards meeting the needs of the world's majority; renewable energy resources need to receive a greater share of attention; natural resources and policies need to be shifted from the arms race to the protection of agronomic and biological resource systems (Brown 1984, 208). All these policy requirements rest on the political will which the contributors to the *Limits to Growth* found so singularly lacking in their Report of fifteen years ago. Before we consider whether the decline of the world economy opens up possibilities for sustainable development that might enhance the importance of environmental values, we need to consider the way in which the issue of growth and sustainability is handled by economic theorists.

Economic growth and world trade

Most development economists and international agencies take the view that without fairly free international trade it is extremely difficult to see how all but the largest developing countries will be able to get sustained economic growth under way. The neo-classical case is that the 'gains from trade' outweigh the losses. However, this approach to trade and development fails to address some serious issues:

1 Even the textbooks on international-trade theory acknowledge that the gains may be very unevenly divided as between countries or trading blocks, depending on their relative economic 'muscle' and their disposition to employ this power.
2 Neo-classical theory also acknowledges that there will be losers as well as gainers. However it does not identify a *process* at work which will 'equalize' these advantages/disadvantages to individual countries. The issue of global inequality is one which orthodox economics can explain but is powerless to address. Few people outside the professional cadres which make up the development agencies fully realize this.
3 Orthodox definitions of 'development' imply that economic growth is broadly beneficial and that freer trade will stimulate growth. However, the assertion that economic growth is beneficial for whole societies can be questioned. Similarly, without costing the environmental and social consequences of growth, it may be a mistake to pursue growth rather than 'development' in a wider sense.

The concentration on 'growth' has served to obscure the fact that resource depletion and unsustainable development are a direct consequence of growth itself. Where agencies like the World Bank have made loans to promote high-growth sectors, the environmental and social consequences have often been disastrous. As demonstrated later in this chapter, the effects of ranching in the humid tropics have usually proved negative in both social and environmental terms. The returns on capital represented by investment in ranching obscure the progressive squandering of a unique global asset – tropical ecosystems and the people who live in harmony with them, which are frequently excluded from calculation. Conventional economic development identifies the optimal conditions for exploiting resources, a rather different process from considering sustainable development. As we have seen the different dimensions of sustainability – social, environmental and ethical – are an essential part of the development debate, but ones which the concern with 'growth' often leads us to miss.

If the existence of international trade in the modern world is essential to economic growth in less developed countries, it has also ensured that natural resources are also exploited for short-term profit. Deteriorating terms of trade for poor countries have contributed to the clearing of rain forests to make way for export-led stock-raising. Similarly a poor balance-of-payments position is likely to contribute to a country's over-dependence on fuelwood, especially given the high costs of petroleum-based energy. Alternatively a fall in price, in relative terms, for key commodity crops usually increases the use of land for these rather than for food crops, and this in turn increases food imports into countries that could be self-sufficient. The point is that decisions over the use to which natural resources are put are clearly influenced, directly and indirectly, by the trading patterns established by the developed countries. Environmental problems are not reducible to international economic relations, but they would not have assumed their present gravity if the developing countries had been able to practise the sustainable-resource methods which often formed part of their traditional systems, as we shall see in chapter 5.

It is within this context, in which sustainable practices have been neglected in favour of closer economic dependence on Northern technology and markets, that we need to understand the current environmental crisis. For example, advising African countries on how to develop their resources has become a major industry, with European and North American consulting firms charging as much as $180,000 for a year of the professional expert's time. More than half the $7–8 billion spent yearly by development donors in Africa goes to finance the 80,000 expatriates working for public agencies under official aid programmes (Timberlake 1985, 8). It may reasonably be asked who is providing development assistance to whom under these circumstances.

Awareness of the serious nature of Africa's environmental crisis has done little to alter the approach of international development agencies. The World Bank's 'Berg Report' (1981) on Africa advocates export-led development as a development priority, rather than goals such as self-sufficiency in food. The solution to Africa's growing external debts, debts like those which have already made a deep impact in Latin America, is held to be the 'East-Asian' model of export-led performance: Taiwan, South Korea, Hong Kong. The Berg Report recommended cutting government subsidies and privatizing the public sector of most African countries with debt problems as a solution to African insolvency. Interestingly, the emphasis in explaining these problems was placed squarely on 'environmental' causes and population increase rather than externally induced structural problems. However, export-led development is a perilous, if not impossible route for most African countries, at a time of high interest rates and worsening international terms of trade. Africa is likely to experience what Latin America already knows: that private banks can only get repayments on their loans if they lend more money to the debtor. The alternative of a debt moratorium is still considered by most countries a much greater risk than defaulting. Most African countries are agreed that food sufficiency should be their continental goal, but they are almost powerless to achieve it while they are saddled with the need to increase export income. How, they ask, can we export more to the developed countries, as the International Monetary Fund (IMF) and the World Bank recommend, when the developed countries are trying to restrict their imports? Faced with the need for 'structural adjustment', they ask the question put by Susan George (1985b):

> If the International Monetary Fund believed (which it patently does not) that economic growth can also result from greater social equality, access to education, health-care and other basic services, fairer income distribution etc., it could make such objectives part of its programmes . . .

At the moment it is quite possible to have economic growth in some less developed countries without 'development'. Close trading relations with the North, as recommended in the Brandt Report, are likely to have negative effects on many less-developed-country economies, partly because they are linked to the policies urged on less-developed-country governments by international agencies like the World Bank and the IMF. These agencies see their primary role as that of guaranteeing the maintenance of the 'trade' process which contributed to the developing countries' indebtedness in the first place. The capacity of the South to mount long-term sustainability is a problem both for the less developed countries which are most 'developed' (the bigger countries in Latin America, and some in East Asia) and those which are least developed (the Sahelian countries, Bangladesh, Haiti). For the first group of semi-industrialized countries, rapid economic growth has meant that they have inherited many of the environmental problems of the 'rich world' in an exaggerated form. For the second group declining per capita incomes and resource exhaustion have led to mass migration, the resettlement of populations, increasing population pressures and frequently financial bankruptcy. In both cases the form of their insertion within the international economy has reduced their capacity to solve their own environmental problems.

The combined effect of the World Bank, the IMF and the United States government borrowing has been to maintain high interest rates, to increase pressure on less-developed-country debtors to export more and consume less, and to starve much of the developing world of much needed capital. The costs of United States government policy are particularly painful in countries like those of sub-Saharan Africa, where more than 60 per cent of export earnings come from commodities for which the price elasticity of demand is such that an increase in export *volume* (urged by the IMF and the Berg Report) would actually reduce export *earnings* (Godfrey 1985, 178). Without altering fundamentally both the way the debt is managed, and the policies of the developed countries towards trade and investment in the South, export-led solutions threaten to put at risk many of the achievements of post-colonial administration.

The solution to these problems is not global economic growth, in the language of the Brandt Report, for the struggle to effect such growth might in itself be damaging to the poor who are already paying the price for structural 'adjustment'. The pursuit of economic growth, unchecked by environmental considerations, can accelerate, among other things, topsoil losses, the scarcity of fresh water, the deterioration of grassland and deforestation. These apparently inexorable processes are a consequence of the policies being advocated today for Africa and Latin America: increase exports, buy in expertise from foreign consultants, construct big dams, bring more land under export-crop production. The irony is that these policies not only have negative environmental consequences, they also frequently fail to meet their economic objectives.

Measured by value or volume there has been *no* growth in world exports in the 1980s, the peak US$1,868 billion having been achieved in 1980 (Brown 1984, 18). The unsustainable utilization of resources for short-term 'growth' objectives might ultimately fail to bring about economic growth itself.

Finally, it is sometimes argued that population pressure is the major obstacle to securing development in the South, as if limiting population could be divorced from the strategies poor people adopt in pursuing their livelihoods. Reference is made to increasing family size preventing 'human needs' being met, when large families are precisely one of the strategies open to the poor to ensure their own survival. What needs to be recognized is the major influence of socio-economic conditions, the *vulnerability* as well as the absolute poverty of people, on the number of children in the family. The conventional argument about population growth could be reversed: efforts to slow down population growth will continue to be frustrated until meeting basic human needs is considered the priority of development.

Aid and the monetization of local economies

The conventional economic wisdom, as we have seen, is that 'freeing trade' promotes economic growth, and with it 'development'. We began by questioning this pattern of causation and asking whether sustainable development did not imply objectives other than economic growth. The environmental effects of liberalizing trade and stimulating agricultural exports have often been harmful and will continue to be harmful until more attention is paid to the ecological and cultural systems which convert 'natural resources' into livelihoods.

Few international issues have focused as much attention in recent years as 'aid', but until recently the discussion of development aid was divorced from a consideration of long-term environmental effects. Apologists for aid, especially food aid, have powerful ideological support in popular assumptions which link helping people in the Third World with general 'neighbourliness'. Of course, the media in developed countries frequently highlight scandals, and aid is seen to have been diverted from its original purpose.

There are many popular misconceptions about aid. Aid, especially bilateral aid between countries, is not directed simply to alleviating hunger and poverty. Even if it has a net effect which is advantageous to the poor, for whom much of it was intended, it is often more beneficial to the better-off in less developed countries. By reducing peoples' capacity to manage their own environment on a sustainable productive basis, aid also contributes to the problem it was intended to address. Food aid, in particular, can increase food dependency and remove responsibility for the environment from those who have most to gain from its beneficial long-term management. This was succinctly expressed by a Senegalese farmer in the following way:

> We don't want to be called bush people. We are citizens like everyone else. We want to be free to sell to whom we choose, and to buy from whom we choose. We too want to set our own prices. No one ought to make us concentrate on cotton: we should grow what we think is best for us. They tell us to follow their programmes and we'll get their assistance, and when we follow that programme the prices of their ploughs double and the prices for our crops stay the same . . . (quoted by Bernard Lecomte in Rose 1985).

Most 'donor' countries, in a position to give aid, give it to countries with which they are politically sympathetic. Egypt, Israel and Turkey between them receive four-fifths of all United States bilateral aid, US$1,900 million in 1980. The same three countries received almost the same amount of US military aid (US$1,800 million in the same year). A few countries receive a significant amount of United States bilateral aid without much military assistance (India, Indonesia and Bangladesh among them) but even this combined total adds up to no more than the United States gives Egypt in military spending, and not much more than half what it gives to Israel (Lappé, Collins, and Kinley 1981). The Eastern Bloc countries for their part give only one-tenth as much aid as the West, and three-quarters of it goes to three beleaguered allies: Vietnam, Cuba and Afghanistan (Harrison 1985). Aid does not simply have political strings: it is clearly important as part of global political strategies. Although aid to Africa has increased as a share of total aid in recent years, the total amount of aid received fell from $8.1 billion in 1980 to $7.7 billion in 1983. This was the period when, once again, large-scale drought began to hit the Sahelian countries and

Ethiopia, while the external-debt problem made itself felt in Africa. It is clear that aid is not given to those who need it most, but to those whom the aid-givers choose for political and strategic reasons, rather than the eradication of poverty.

Aid also distinguishes unfairly between those who receive it. Ethiopia has the lowest per capita income in Africa (only $140 in 1983) but received a mere $8 per capita in aid. Neighbouring Somalia, where incomes per head are twice as high, was rewarded for her friendship with the West by ten times as much aid per capita in 1984. There is evidence too that cash handouts tend to reach those farmers who are well connected, rather than those in greatest need, reinforcing public scepticism in developed countries, when it is exposed.

The consequences of food aid, for people and their environment, are often more severe. Oxfam concluded, in a recent publication (Jackson and Eade 1982, 91), that:

> . . . there are inherent problems associated with food aid which are peculiar to it and make it a particularly cumbersome and inappropriate means of providing assistance . . . because of its appeal at a simple level . . . food aid's inherent weaknesses have been largely overlooked.

Some of the reasons for which food aid exists have little to do with development, and even less to do with sustainable development. For example, the United States was heavily engaged in its PL480 programme when it was a large surplus producer of grains. The European Common Market is today involved in food-aid programmes to Africa as a consequence of its own agricultural surplus problem. Food and agricultural policy in the EEC was never designed with surpluses in mind, but, perhaps more seriously, it was not designed with famine relief in mind either. Most official European food aid has arrived too late, when the food insecurity of the African countries concerned could have been anticipated years in advance.

The food surpluses (and deficits) created in the developed countries also dictate policy towards the growth of agro-industry in the South. During much of the 1960s and 1970s the United States experienced problems with grain surpluses, which had either to be stockpiled or exported (see table 4.2). The solution that was found to this problem was the sale of grains at subsidized prices, notably to

Table 4.2 The changing pattern of the world grain trade 1950–83 (million tonnes)

Region	1950*	1960	1970	1980	1983
North America	+23	+39	+56	+131	+122
Latin America	+ 1	0	+ 4	− 10	− 3
Western Europe	−22	−25	−30	− 16	+ 2
Eastern Europe/USSR	0	0	0	− 46	− 39
Africa	0	− 2	− 5	− 15	− 20
Asia	− 6	−17	−37	− 63	− 71
Australia/New Zealand	+ 3	+ 6	+12	+ 19	+ 9

Notes:
Plus sign indicates net exports: minus sign net imports
* Average for 1948–52
Sources: FAO *Production Yearbook* (Rome, various years); US Department of Agriculture, *Foreign Agriculture Circular* (August 1983); compiled by Worldwatch

the Soviet Union, and the dumping of surpluses in less developed countries which were still net deficit producers, like India. The PL480 programme, for example, enabled successive United States governments to disguise domestic market instability behind the appearance of development aid. Once surpluses had been disposed of, American farmers reduced their acreages again, and the system was 'stabilized'. In the early 1970s the United States experienced an enormous drop in total acreage devoted to grains – from 120 million acres to 81 million acres at one point in the expansion/contraction cycle. Most less developed countries, for their part, could not afford to buy grain from abroad until the United States' stocks were in surplus and the price had come down. Food thus became a very effective weapon of international diplomacy, evidenced by the refusal of the United States to provide the Bangladesh government with food aid in 1974 when the Bangladesh famine was at its height. (This move was prompted by the fact that the Bangladeshis were selling their jute to Cuba.) Again, in September 1981, President Reagan imposed a wheat embargo on Nicaragua as part of his policy to destabilize the Sandinista administration.

Food aid, like much aid in general, helps strengthen the position of urban workers and urban élites at the expense of rural groups (Lipton 1977, Singer and Ansari 1977). By making cheap imported food available it also acts as a disincentive to local farmers to produce. Where they can, they avail themselves of food aid (Hayter and Watson 1985). It also, and this is a critical issue, changes the diet of the poor, introducing them to highly valued imports like wheat bread, at the expense of indigenous food grains and root crops, millet, sorghum and cassava. The shift in tastes reduces demand for the local farmers' produce which in turn holds down the prices local farmers receive. The neglect of basic agricultural research into the crops grown by small farmers, one of the abiding weaknesses of African development policy, completes the vicious circle of poverty.

The Green Revolution in Asia was a dramatic example of the way that the production of food grains could be increased through improved seed varieties and heavy dependence on chemical fertilizers and pesticides. The increases in aggregate production which followed enabled some countries, like India, to build up national buffer stocks of food staples, ensuring that they do not experience shortfalls like those of 1971 when over one million died because food was not available. However, the ecological effects of this Revolution are being counted today, almost two decades after it got under way. They include the salinization of irrigation water, the polluting effects of chemical sprays and the increasing resistance of pest species to insecticides. The total environmental cost is even greater, and account should be taken of the narrowing of popular diets, the increased vulnerability of small farmers to indebtedness and the external dependence of oil-importing countries (like India) on outside suppliers. The Green Revolution has proved to be a mixed blessing.

Most of the technical assistance that has gone into African and Latin American agriculture has been directed at producing cash crops for the international market. The Green Revolution has mixed effects, but it was at least targeted at food crops. Cash crops earn foreign exchange to pay for capital goods (Africa), intermediate goods (Latin America) and oil. Increasingly cash crops pay off the interest on the external debt which was itself incurred largely through the acquisition of prestigious and capital-intensive development infrastructure. This is another vicious circle, made up not of the producers and consumers of food, but the producers and consumers of transnational technology. The contrasts between Latin America/Africa and many parts of Asia are vivid. Inevitably Africa has become a food-dependent continent as a result. Cereal imports rose to 32 million tons in 1984, when African farmers themselves were able to produce less than two-thirds of the continent's cereal needs. According to the Food and Agricultural Organization, Africa will be growing just over one half of its own cereal needs by the year 2000.

Cash crops have also contributed to the monetization of local economies. By 'local economies' is meant the system of production and distribution that local people manage themselves. Within local economies the labour of women, in particular, is as undervalued in financial terms as it is essential in livelihood terms. With the impact of market forces women's labour is taken for granted, and the labour process is transformed, to the benefit of men. Women's responsibility for reproduction as well as production places them in a disadvantageous position in relation to new market opportunities. It is women who nurture the children, feed the family and provide much of the 'casual' paid labour which underpins commodity production for the market. It is also women, critically, who interact most closely with the natural environment: collecting the fuelwood, carrying the household's water long distances, tending the vegetable garden. Women therefore bear the brunt of environmental degradation, through their proximity and dependence upon the environment, while *also* being held responsible for this decline. Unable to reverse the erosion of resources to which the household has access, women are placed in the impossible position of acting as guardians of an environment which is as undervalued and exploited as their own labour.

The effects of an increase in commodity production for the market and the family's dependence on wage labour fall disproportionately on women and, through them, the environment. Male involvement in the monetized sector is invariably greater than that of women across a range of rural societies. Similarly, women bear major responsibility for intensive non-mechanized production at one end of the food system, the end which is casually ignored by large-scale development projects. At the 'other' end of the food system it is usually women who look after food-processing, storage and family nutrition, activities that are increasingly displaced from the household by the 'development' process and relocated in commercial, agro-industrial plants controlled by men. It is essential to recognize that the loss of primary environmental control by rural households is bound up with the marginal status and declining economic power of women. This is discussed further in chapter 6.

What are the specifically *environmental* effects of development aid?

Can we make connections between the economic and social costs of large-scale development planning and the disintegration of the rural environment in so much of the South? The answer is that we can but that the connections are often obscured by the superficial attraction of large projects. As Bruce Rich (1985, 56) reports:

> Projects involving the spraying of vast areas with massive amounts of hard pesticides that are often banned in Western countries, projects involving the construction of big hydropower and perennial irrigation schemes which require the removal of large populations on to infertile land, with no proper compensation, and which must inevitably lead to widespread waterlogging and soil salinization as well as the spread of waterborne diseases.

In 1983 the four Multilateral Development Banks (World Bank, Inter American Development Bank, Asian Development Bank and African Development Bank) allocated over half their project loans to work in environmentally sensitive areas, such as agriculture, rural development, dam and irrigation schemes, but employed an absurdly small staff to oversee the 'environmental consequences' of their work. As we shall see in chapter 7, the Office of Environmental and Scientific Affairs of the World Bank employs a minuscule staff to review the prospective environmental impacts of most of the Bank's 315 projects. Evidence from within this office is eloquent testimony to the cursory way in which environmental impacts are treated (Watson 1985 in Hayter and Watson 1985).

In sub-Saharan Africa aid has largely been directed towards projects in which medium- and long-term environmental effects have scarcely been considered. In most of the Sahelian countries (Chad, Niger, Mali, Burkina Faso, Mauritania, Senegal, Gambia and

Table 4.3 Human numbers affected by desertification (millions)

	1977	*1984*	*2000*
Populations at risk of desertification	650	850	1200
Rural component	280	500	600
Severely affected	80	230	
Rural component of severely affected	57	135	

Source: Compiled from UNEP, *General Assessment of Progress in the Implementation of the Plan of Action to Combat Desertification 1978–84* (Nairobi 1984)

Cape Verde) population has been growing at a steady 2.5 per cent a year, while food production has grown at only 1 per cent a year. The natural environment is particularly vulnerable under these circumstances, as even small numbers of people can have a large impact on fragile environments and it is hard to increase productivity on poor soils (see table 4.3). A decade after the Sahelian famine of the early 1970s 'aid money has had little effect' (Tinker 1983, 9). Deforestation, overgrazing and over-cultivation are not simply the unpleasant consequences of food and energy shortages, they are one of the results of considering development assistance in terms of short-term production gains rather than longer-term sustainability.

First, reforestation has received only a small proportion of the development aid provided to the Sahelian countries and only 1.4 per cent of the aid budget has gone towards forestry schemes managed by local people themselves. In the urban areas most of the people cook with wood or charcoal, as kerosene is too expensive for all but the relatively rich. The various attempts to introduce eucalyptus plantations in parts of Karnataka and other areas of India are clear examples of the disadvantages of a cash crop and the commercialization of forestry. Work by Vandana Shiva and others shows how the firewood crisis will actually worsen despite the present growth of eucalyptus plantations (Shiva 1986).

Second, the objective of most agencies working in the Sahel has been to sedentarize the nomadic pastoralists of the region, by providing them with incentives to 'off-take' some of their animals for meat, while maintaining herds of modest size for milk production. This strategy has been undertaken in the face of increasing evidence that this is not what pastoralists want, nor does it correspond with the way they are likely to act. In theory the pastoralists' growing involvement with the cash economy, 'their need for ready money to buy transistors and tents and sugar and sunglasses, will gradually persuade them to increase their off-take' of young animals (Tinker 1983). In practice those pastoralists who have survived are rightly suspicious of governments who are intent on fundamentally altering their livelihoods. The hope that many of them will settle near boreholes has driven them to near-extinction, while they have found increasing difficulty in selling their animals to buy the food crops they need.

As we have seen small farmers are equally exposed to the exigencies of the market. Agriculture in the Sahel is only really possible where the average rainfall exceeds 400 mm a year. It also depends critically on adequate fallow periods. The drive to monetize local economies by increasing the marketed surplus and shifting to cash crops has forced the fallow periods within narrower limits, or they have been abandoned entirely. Similarly, under environmental pressures that cannot be sustained, farmers push northwards into the pastoral zone, tilling soil which should be under grass, as it is too arid for permanent crop production. Yields have fallen as a result: the area under cultivation in Niger has increased by almost a half, but production is only 4 per cent more and production per hectare three-quarters what it was in 1969 (Tinker 1983, 11). Such is the despair among aid agencies at the advance of the desert in the north of Burkina Faso that the World Bank circulated a private report recently suggesting that the time had come to cease major investment in the parched north of the country and, instead, encourage people to migrate to the south to lands newly cleansed of diseases such as river blindness (*New York Times*, 29 November 1984).

There is also evidence that the interruption of the pastoralists' cycle, which has forced nomadic peoples into overgrazing land, stripping it bare of vegetation, also has climatic effects. According to one recent authoritative paper, published in the *Canadian Journal of Zoology* (1985) the traditional pastoralists' system, interrupted by 'well-intentioned aid agencies' in favour of the expansion of peanut production, was the only way in which humidity could be retained in the soils. Migratory cattle herds, deprived of winter grazing land, have been encouraged to settle by the provision of water holes on winter rangeland. The pastoralists obliged, so cutting off their cattle from the northern midsummer grasslands that had been essential to them. They were also attracted by the medical services being provided for them in the areas of settlement. The stage was now set for disaster: more cattle compressed into a smaller area, grasses grazed throughout the year without respite, human population increasing fast thanks to improved medical care. The result was famine conditions.

This environmental degradation cannot be reversed. Land under vegetation absorbs more heat than bare sandy soil. Thus, when there is plant cover, the ground is warmer. Consequently there are more vigorous thermal undercurrents which take moisture, provided by that very plant cover, up to high altitudes where it condenses.

Overgrazing strips vegetation and leads to a new stable state of bare cool soil, lower rainfall and sparse vegetation. According to a recent report in *The Times* (18 October 1985) 'although the verdict is not wholly certain it is possible that parts of the western Sahel have already switched to a new lower rainfall regime'. Environmental factors are important, not because economic policy cannot change them, but because it does.

Debt and interest rates

It is difficult to exaggerate the magnitude of the debt crisis. In 1986 Mexico, which earns more in foreign exchange from oil than all its other exports put together, is expected to spend its entire revenue from oil in paying off the interest on its external debt. The Mexican government also borrows at home and pays interest on this domestic debt. In 1985 almost *two-thirds* of the Mexican government's income was spent paying off this domestic debt, leaving just over a third to meet total public expenditure. As Mexico has become more indebted, so the IMF and the international banks, which are its creditors, insist that the government open up the Mexican economy to foreign investment, sell off publicly owned companies and reduce subsidies, especially on food. The last measure is particularly perverse in its effects: as subsidies are withdrawn, inflation increases, interest rates rise further and the Mexican government pays more to finance its debt.

The situation in Africa is less severe in terms of the absolute magnitude of the debt, but in relative terms it is probably worse. In 1983 there were no African countries among the big debtors. Today the total external debt of the 42 sub-Saharan economies is in the order of US$130–135 billion. The true average debt-service-to-exports ratio, excluding debt arrears and reschedulings, in these countries was about 35 per cent in 1985 (Green and Griffith-Jones 1985). This is less than in many Latin American countries – Brazil, Mexico, Chile and Argentina are spending most of their export earnings on debt interest – but it is critical for another reason. Put starkly, much of sub-Saharan Africa is literally starving while debt interest is being paid.

The case of the Sudan is illustrative of what is happening in Africa. The Sudan has a US$78 million proposal for emergency aid and US$213 million in interest due, after rescheduling on US$10 billion of external debts. Paying interest at this rate, for a poor country like the Sudan, means effectively cutting grain imports needed to avert starvation or fuel imports needed to keep the economy operating. In the words of a recent report, a failure to alter the existing debt imbalance will mean 'in human terms for most Africans . . . increased political and social instability (or worse), severe deprivation for a majority – and premature death for hundreds of thousands a year' (Green and Griffith-Jones 1985, 221).

It is worth remembering how the debt crisis arose in the first place. After the first oil shock in 1973/4 there was a fourfold increase in oil prices. The OPEC countries invested their large export surpluses in the Euro-currency market and the international banks. At that time conditions for borrowing were easy and the banks *wanted* to lend. It was also profitable to borrow because interest rates were low compared with world inflation figures. Interest rates were often negative in real terms. World inflation figures also cut the value of the accumulated debt. The banks then set about lending money enabling developing countries to 'adjust' to the effects of oil price rises. 'Adjustment', like so much else in the international development 'doublespeak', is often a euphemism for policies which imply a drop in living standards.

The real crisis came after the second oil shock in 1979/80. Just when debtor countries needed credit, they found high real-interest rates, and the availability of credit from OPEC also declined. By August 1982 the Mexican debt crisis forced the IMF, the United States government and the private banks to assemble a huge debt-rescue package. Other countries followed, in Latin America and elsewhere. Accompanying the increased debts was an overvalued dollar-exchange rate which undermined national currencies and led to the flight of capital: over US$11 billion dollars left Argentina in 1981/2; US$17 billion left Mexico in the same period.

Given the sacrifices implied by structural adjustment policy, it may well be asked what prevents debtor countries defaulting completely? If payments on the debt are suspended (moratorium), the costs to the 30 or so international banks which have benefited from high interest rates would be serious. It would have cost them US$14 billion in 1983, had payments been suspended, making a considerable impact on United States capital markets. If there were a debt default, the United States banks involved would be ruined. It is not surprising, then, that the banks (and the IMF) favour the idea of debt rescheduling. This brings them rescheduling fees, it spreads the interest rate and it protects their assets at book value.

The short-term solutions that have been found to the debt crisis are obviously good for the banks, but the more critical question is what they imply for the debtors. O'Brien argues that Latin American countries, at least, should seriously consider the advantages of a co-ordinated default (O'Brien 1986). The way that you interpret a managed solution to the debt depends on what you think are the prospects for the world economy. An objective assessment would hardly lead to optimism. The United States is the best example of a rich country engaged in deficit financing to maintain consumption levels. In 1985 Susan George commented on the fact that the United States, with a population of 220 million owed nearly US$6,000 billion, approximately twice as much debt as all the Third World countries put together with one-twelfth the number of people (George 1985b).

One of the most dramatic effects of the United States government's borrowing activities, combined with high interest rates, is the American 'farm crisis'. In the popular imagination the American family farmer, usually depicted as a bastion of democracy and friend to the environment, is being squeezed out by rapacious money-lenders. This picture leaves out as much as it depicts. The staggering debt load of United States farmers (US$18 billion in 1985 compared with total farm income of US$25 billion) is partly a consequence of the fiscal crisis, and the Reagan administration's borrowing. It is also, like the plight of the less developed countries, a result of borrowing in times of plenty against (in the farmers' case) an appreciating asset – the land. As land prices dropped, and interest rates rose, more farmers found themselves in debt to the banks. They were also trapped in a cost–price squeeze, having become increasingly dependent on industrial inputs to farming. In 1985 one in twenty American farmers left farming, an increase over the average annual rate of 3–4 per cent, in recent years. The asset values of those that remained declined by 40 per cent. American farmers are already experiencing what many of their West European cousins still fear, as they are unable to repay their debts.

The remarkable thing about the United States farm crisis is that it has been discussed almost exclusively in fiscal terms. The roots go much deeper and suggest that increased military expenditure is an important factor. Under President Carter military expenditure was

$160 billion p.a.: in 1985 under President Reagan it was $400 billion. The United States' economy is increasingly dependent on arms production, accounting for over one-third of all research and development in science and technology, and two-thirds of public expenditure in these fields.

The farm crisis in North America should also give us cause to examine the relationship between increasingly intensive agriculture and farmer indebtedness. As the agro-industrial complex expands, and farming becomes dependent upon industrial inputs, indebtedness and bankruptcies increase. Modern agriculture does not remove the threat of insolvency, it makes it more likely. This should lead us to examine the nature of indebtedness within the context of environmental sustainability. From the personal level to the national and international levels, increased indebtedness assumes a continual growth of personal income and GNP. Indebtedness is not a curiosity of bad economic management, it is a symptom of a deeper malaise that equates 'development' with the conversion of natural resources into consumable products, many of which we produce but cannot sell. Indebtedness makes claims on the environment which are unsupportable and unsustainable. It is part of the motor of destruction that we see both in the developed world and in less developed countries, and to which an alternative is urgently required.

Most debtor countries in the South are currently at the mercy of a logic which has nothing to do with meeting basic needs or reducing the vulnerability of the poor. The IMF, as Susan George has pointed out, exists to promote the growth and development of world trade (George 1985b). World trade, as we have seen, can be expanded without net benefits accruing to poorer trading 'partners'. The IMF exists to ensure that trade and economic growth are maintained, whatever the social and environmental consequences. Countries were once considered to be in trouble when the average debt ratio exceeded 25 per cent of their export earnings, and this makes it absolutely clear just how much trouble they are in now that many countries in Africa have a debt ratio nearer to 35 per cent. What is required is not a renegotiation of their debts, and a rescheduling of their indebtedness, but a completely new approach to the debt problem. Such an approach will need to address the vexed issue of national sovereignty and the operations of transnational corporations.

Sovereignty and the multinationals

During the last decade there has been mounting criticism of the role of transnational corporations (TNCs) in the 'development' of the South. Most of this criticism has focused on the 'irresponsibility' of TNCs which put profits before nature conservation and, for that matter, human livelihoods. Criticism tends to overlook the essentially rational and systematic way in which TNCs operate, usually with the support or complicity of national governments in less developed countries. It also overlooks the fact that TNCs act as self-contained systems, commissioning and utilizing technology, creating and exploiting consumer markets in the North, using 'transfer pricing' to ensure that the benefits of peripheral development accrue to the centre. The TNCs are the main beneficiaries of a free-trade/open-market system; they benefit from being able to sell expensively and buy cheaply, whether it is labour, primary products or technology.

There are also political advantages to be gained from commanding the process of resource–food conversion without necessarily having a major stake in the ownership of another country's resources. Nationalist sentiment was a potent force in persuading the Brazilian government to tighten its control over foreign-owned corporations in the development of the Amazon, although the model of 'development' within which these corporations operated has still not been seriously challenged (Branford and Glock 1985). Indeed the state is frequently the guarantor of TNCs, not only by providing a legal basis for their operation, but by underwriting them financially. If we look at figures for the development of Latin American agriculture in the 1970s (Lopez Cordovez 1982) we see that chemical-fertilizer use rose by 8.5 per cent *annually*, and the use of pesticides by 8.4 per cent *annually*. Tractorization was also an expanding area, experiencing a 4.8-per-cent annual growth rate. During the 1970s much of Latin America's agriculture became increasingly tied to high-technology inputs which received the bulk of state subsidies and preferential credits. In 1975 over half of Latin America's chemical fertilizers were imported. The interest of United States' companies in tractor production was similarly important. The major thrust of agricultural policy in Latin America was to ensure that modern, commercial agriculture could be undertaken 'competitively' through subsidized credit, fiscal concessions to importers and industrial suppliers. Guaranteed prices were paid to co-operating farmers and scarce public investment was allocated to large-scale physical infrastructure in areas like NE Brazil and NW Mexico, where heavily capitalized, irrigated agriculture was promoted. These policies met the demands of external agencies for 'modernization' of agriculture, while providing opportunities for the realization of profits at critical stages in the production cycle.

The increasing activities of TNCs in the South is thus the outcome of both a concern with their domestic profitability, and changes in the technology that TNCs employ. TNCs have themselves affected the international division of labour. Third World agricultural producers, notably in Latin America where agriculture has long been dominated by North American interests, have come to play a provisioning role within an evolving economic and technological system which has its origin and rationale in the North.

Another essential element in the way TNCs operate is their need to gain access to raw materials and labour. In their search to find supplies of both, TNCs inevitably act in ways which are prejudicial to the economic and political sovereignty of national states. Agro-industrial operations in California and Mexico, such as those described by Burbach and Flynn (1980) are thus variations on the same theme. In California, American companies own the land used in agribusiness operations and the capital to exploit it. They need to 'import' the Mexican labour employed in the fields and the packing plants. In Mexico the land used by United States agribusiness is Mexican-owned, and the labour is freely available from the impoverished smallholder sector, but the capital has to be 'imported' from the United States. The frontier is a flexible concept which corresponds more closely with geography than with the workings of agro-industry. TNCs increasingly make decisions that transcend their basis in legal ownership and materially affect nutrition, land tenure and the environment.

By commanding the necessary technology TNCs can also ensure that there is an international division of labour in another sense: not only do less developed countries provide most of the raw materials for industrial processing carried out in the North, but by employing peasant farmers on a 'contractual' basis TNCs also enable the costs of maintaining the labour force to be passed on to poorer households. For most of these people employment by TNCs is on a casual basis. Whether they produce on their own land or work as seasonal labourers, they need other sources of livelihood support, from the land or from urban employment. Many are being forced into making ever increasing demands on their environment by their dependence on the market. They resort to short-term unsustainable agronomic practices, shortening fallows, acquiring fertilizers on credit, removing tree and plant cover, reducing the range of crops within their 'farming system'. Others become 'ecological refugees', forsaking their homes for other areas where they can establish settlements. A substantial part of transnational migration from Haiti, the Sahel and Central America has environmental, as well as political, roots.

This underlines another 'knock-on' effect of transnational penetration. Many of the consequences of agricultural intensification are directly attributable to agribusiness penetration. Agribusiness implies agricultural intensification, specialization in a few laboratory-bred crops and their treatment with chemical fertilizers. The problem is that, even when the target of the production system (on-farm productivity) behaves as expected, other elements in the farmer's system are usually adversely affected (other crops, animals, other income-earning opportunities), while increasing intensification brings deterioration to elements *outside* the farming system (through pesticide use, land erosion and changes in diet, for example). TNCs, in making themselves more competitive, reduce the flexibility and risk-avoidance which are essential to successful 'peasant' agriculture, as we shall see in chapter 7.

Perhaps the most vivid example of transnational penetration is the destruction of the tropical rainforests. The statistics are appalling: in the last 25 years more than a quarter of Central America's rainforest has been turned to grass, and almost all the beef produced on it has gone to American hamburger chains. In 1960 the United States hardly imported beef at all. By 1981 800,000 tons were being imported annually at a price less than half that obtained in the United States. As a recent article in the journal *Environment* put it: 'the average domestic cat in the United States now consumes more beef than the average Central American' (Nations and Korner 1983). The 1970s was the decade in which one product received more development aid than any other in human history: the promotion of beef production in Latin America attracted over US$10 billion from the World Bank and the Inter American Development Bank, at 1983 prices (Rich 1985).

Elsewhere in the humid tropics the story has been much the same: large-scale finance for forest clearing and resettlement schemes, with the accent in Africa and Asia on timber production rather than ranching. In Indonesia it is planned to 'resettle' four million people from Java, Bali and Lombok on the 'Outer Islands', which have little physical infrastructure and few support services. The human cost cannot be counted in full, since there is nobody but the settlers to count it. Those who have participated in some of the transmigration schemes have returned to speak about the experience in a tone not dissimilar to that of survivors from the holocaust. As demonstrated by evidence to the Brundtland Commission, given in Jakarta in 1985, transmigration is neither a substitute for redistributive land reform nor a viable solution to the over-population of Java (Brundtland 1985b).

Tropical forests are destroyed primarily for economic reasons and, although it is important that there is growing awareness of the ecological problems produced, such awareness alone cannot be expected to turn the tide. Only radical changes in structural policies can do that. At the present time eight LDCs earn between them US$100 million a year from timber exports (Brundtland 1985a, 24). Earning foreign exchange on this scale is not merely an objective of governments in the South, as we saw above, it is also the objective of their creditors, the IMF and the international banking community.

The need to earn foreign exchange is compounded by the need to grow more food, and many countries are less exercised about what food is produced than that it is produced at all. The wilderness areas whose conservation even the World Bank now recognizes (Goodland 1985) represent wasted, unproductive land to financially bankrupt governments in the South, intent on avoiding radical land reform at any price. The colonization of the tropical forest thus becomes a national goal, invested with the kind of symbolic importance which ecologically minded people reserve for endangered species. It is important to recognize these contradictions, painful though they are, for otherwise we cannot hope to enlist governments, international agencies and private individuals in sustainable tropical-forest utilization.

Enormous economic incentives are offered those who are prepared to play their part in the great 'national' campaigns to bring 'civilization' to the tropical forests. In 1979 incentives to cattle ranching in Brazil cost the Brazilian government US$63,000 for *each* ranching job created (Skinner 1985). Yet figures like these are seldom set against the 'value' to Brazil of exploiting the forests to earn foreign exchange. They are the 'invisible' exports.

In the Mexican humid tropics almost 90 per cent of cattle ranches were owned by private ranching interests in the 1970s. Since 1937 a presidential decree (passed with the authority of President Cárdenas, one of Mexico's more radical leaders) protected between six and nine million hectares of cattle land from the land-reform programme. At that time ranching was most important in the semi-arid north of Mexico. Today it is the tropical south that is the key area in the expansion of ranching, as the forest is cleared. Between 1970 and 1979 grazing land in tropical SE Mexico increased in area by 157 per cent. Between 1950 and 1970 the land under tropical forest in the south, the Lacandona, had already been halved. Today each head of cattle occupies just over 2 hectares of land in Chiapas, land on which stable farming systems could be developed for the thousands of landless in the state. Most of the agricultural credit for this development (57 per cent of it) comes directly from the Mexican government, in the form of short-term loans to buy livestock, capital 'on the hoof'. Mexico's recent brief flirtation with food policy, the Mexican Food System (SAM) caused a few eyebrows to be raised at the scale of government assistance to ranchers. The ranchers themselves, and their supporters among the provincial governing élite, were not outflanked, however. Under the SAM, Chiapas ranchers were asked to devote a quarter of their rangeland to basic grains, particularly maize. This they agreed to do, provided that their ranching interests were left unaffected. This provided for an agreement between the cattlemen and the State which had binding implications for the whole land-reform process in Mexico, since, under the Mexican constitution, the ranchers were only supposed to raise cattle on their land because it was 'unsuitable' for maize production. Since 1982, and the abandonment of a national food policy, Mexican government attempts to gain control of land use in tropical areas have largely evaporated. In the cities the withdrawal of subsidies from 'basic foods' has exposed the urban poor to greater nutritional vulnerability.

It is tempting to regard such stories as examples of the 'careless technology', referred to by Taghi Farvar in his seminal book, written over 15 years ago (Farvar 1970). But 'careless' as the technology may be, human purposes were far from careless. They arise from a short-sighted but, within the prevailing system of rewards, rational attempt to impose a social and technological system on an area quite unsuited to it. The key characteristics of humid tropical ecosystems include incredible species diversity, a highly specialized system of nutrient recycling, uncertain succession responses in the biomass and rapid rates of growth of the biomass. As Norgaard (1984a) has shown, a compatible social system would emphasize a multiple-product, regional, near-subsistence economy, the participation of native people using indigenous knowledge, technologies that evolved in the tropics, together with formal and informal risk sharing. Instead, the system that is implanted in Mexico, as in Brazil, produces a few crops for distant markets, transplants peasants from other parts of the country, uses temperate-zone technology and concentrates decision-making power in the centre, while passing on the risks to the colonizers. It should come as no surprise to learn that under these circumstances so many colonizing families revert to multiple cropping combined with hunting and gathering: the livelihood strategy of the rapidly extinguished indigenous population.

This chapter has argued that the structure of the international economy is partly responsible for the worsening condition of local environments in many parts of the South. The pressure to achieve more economic growth, orientated to external demands in a period of indebtedness, had served to deepen the crisis afflicting the local economy in many rural areas. Instead of the sustainable development of their resources, especially those controlled by women, the strategies of survival forced upon rural households have led to over-intensive cultivation, the depletion of capital stocks (including animals) and migratory patterns designed to increase cash income. Food production on the farm, which in Africa especially is often the province of women, has taken second place to short-term unsustainable strategies which serve to divest the farm household of entrepreneurial control. In the next chapter these questions are addressed historically, as well as geographically, in the role agriculture has played within global patterns of accumulation.

5

The internationalization of the environment

The 'environment' is usually looked upon as located outside ourselves; it is the space that we inhabit. This 'bounded' quality of the environment is seen as its defining characteristic. In this chapter a rather different view is expressed: the 'environment' is looked upon as process rather than form, as the result of a set of relationships between physical space, natural resources and a constantly changing pattern of economic forces. The environment in the international economy is an internationalized environment and one which often exists to serve economic and political interests far removed from a specific physical 'location'. There is, of course, a history to the process through which the environment has been internationalized. Beginning in most cases with the impact of colonial powers, patterns of resource exploitation emerged which enabled these powers to accumulate capital and assume political hegemony over the subordinate 'colony'. At the same time, these patterns of resource exploitation incorporated new products and technologies from other areas of the globe. The development of global agriculture dates from the earliest colonial impact, although it finds its most mature expression in the new plant varieties promoted during the 1960s and 1970s by the international agricultural centres and, most recently, in international biotechnology.

Just as resources, genetic materials and technology have become internationalized in different ways, so the social formations of developing countries have evolved a variety of forms. Some societies have developed out of the 'plantation' mode established in the seventeenth and eighteenth centuries. Other societies have responded to centrifugal rather than centripetal forces, as the colonial ruling classes assumed more control over their affairs and internal markets developed, stimulated by urban growth. Imposing typologies on historically discrete societies is a hazardous venture, but a clear distinction exists between former settler societies, which have continued to develop around primary products for the world market (countries as different as Australia and Ghana), and those in which rapid population growth has accompanied a significant level of industrialization (most of Latin America and East Asia). Clearly we cannot interpret the development trajectory of whole nations on the basis of a theory of international exchange alone, but at the same time it would be false to depict the process of environmental change as separable from international economic forces.

Finally, this chapter examines the way in which attempts to 'manage' the environment internationally, through designated 'biosphere reserves' in vulnerable areas and international agreements to restrict access to 'the commons', have proved largely unsuccessful. As the international division of labour has assumed ever greater importance in the way natural resources are exploited, so the powerful interests behind this process (transnational companies, consumer interests and nation states) seek new technologies to reduce risk and further their social and economic objectives. The interest in conservation is increasingly brought into conflict with the drive to harness nature for marketable, productive ends. The corollary is that environmental management, to be successful, needs to operate through the social groups which are being marginalized by the development process. This contradiction, central to the argument of this book, is explored further in chapters 7 and 8.

The development of global agriculture

The internationalization of agriculture developed around the movement of capital and labour, land being the one factor of production which remained immobile. However, whereas land itself was immobile, the biological resources that depended on land – plant and animal life – were elements in international exchanges, in this respect not unlike capital and labour. Furthermore the technologies which were developed to exploit land-based resources have frequently given way to technologies which alter the very nature of biotic resources themselves, from being environmentally compatible they have become increasingly independent of environmental factors, a true extension of the industrial process.

Before the rise of West European capitalism the international trade in agricultural products was centred on the East Indies, and in particular the island of Malacca. This trade was largely dominated by Arabs, Indians and Chinese and had developed independently of the rise of the European maritime nations (Crow and Thomas, 1982). The object of the European powers – Britain, Holland, Portugal and France – was to control the existing spice trade, in this way guaranteeing that the spices reached the tables of their own ruling classes, for whom condiments were an essential feature in preserving the condition of fresh food.

The conquest of the New World took this process one stage further. The Spanish and Portuguese had plundered Central and South America in a search for gold and silver; agriculture was, at first, almost an incidental activity. However, the labour that was employed in a colonial mining economy had to be fed, and a *criollo* population developed with its own needs. Nevertheless this 'internal' market for food crops was slow to develop. The products which were developed by plantation agriculture were much more important and largely superseded precious metals as the main extractive activity of early colonialism in the Americas. Sugar, in particular, was located in the Caribbean and Brazil where plantations employed large numbers of slaves from Africa. By the end of the nineteenth century approximately nine million African slaves had been forcibly transported to the Americas, six million of them in the eighteenth century alone. Human labour – that of the slaves – acquired 'value' in its own right; plantation production was impossible without a large labour force. However, plantation agriculture existed without slavery in some instances, and products such as rubber, sisal and chinchona (for quinine) were grown under a variety of different labour processes.

In the case of the classic plantation crops labour was transported to the environment in which production took place: slaves, indentured workers or captive indigenous peoples. There was also a 'return' migration, however. Crops such as potatoes and maize were brought from the Americas to Europe, where they quickly established themselves as important staple food crops. In the sixteenth century potatoes yielded four times the carbohydrate content per hectare of the European cereal crops, wheat, oats, rye and barley. Potatoes and maize were brought to Western Europe at about the same time and gradually served to transform the productivity of European agriculture. On the periphery of Europe – Ireland, Poland, Russia and

Scandinavia – the new crops helped to stimulate an agricultural surplus production which was absorbed by the richer, central areas. Increased agricultural productivity in the richer lowland areas of Europe was, in turn, a significant factor in the Industrial Revolution, which was to transform the European landscape in the late eighteenth and nineteenth centuries. The transfer of agricultural products to temperate zones played an essential part in the rise of the industrial nations, whose early mercantilist adventures had opened up the colonies. Similarly new staple crops were transferred from the Americas to Africa: manioc (cassava), sweet potatoes and maize were all brought to Africa by slave ships returning to collect more human cargo. In time they were to become food staples on the African continent, grown by peasants for their own consumption.

The attempts to 'liberalize' the trade in tropical foodstuffs gradually took on a momentum of its own. Many of the products which grew to prominence in the eighteenth and nineteenth centuries in Europe – tea, coffee and cocoa (as well as opium) – were stimulants or narcotics which contributed to new consumer preferences and styles. Recreational activity for much larger numbers of people was defined by the tea-shop or the coffee-house, and medical science made greater use of narcotics in the treatment of physical and mental illnesses. In an important sense the life-styles of the rich, industrialized nations were forged from the convergence of three factors in the global development of agriculture: the movement of crops, the control over trade and the (usually) forcible relocation of labour. Just as consumer preferences in the North help determine present-day forms of resource exploitation in the South so, in the eighteenth and nineteenth centuries, the environment in the South was being restructured around changing social patterns and behaviour in the industrializing countries.

The first characteristic of plantation agriculture, as we have seen, was that it brought together plentiful, cheap labour and the intensive cultivation of one crop. Island economies, which proved particularly suitable for monocrop cultivation, became transformed in the process of supplying northern markets. Between the mid seventeenth century and the late nineteenth century islands such as Ceylon, Java and Cuba were pushed through a cycle of monocrop cultivation. In the case of Ceylon this was cocoa, coffee, chinchona, tea and rubber. In Java cocoa also gave way to coffee, tea, chinchona and rubber. Tobacco was established first in Cuba, followed by sugar, cotton, coffee and sugar again. The shifts from one crop to another in the cycle are attributable to a number of factors. Among the most important was the competition from rival island economies and the heightened susceptibilities of specialized plantation agriculture to pests and disease. The transfer of crops between different islands was testimony to the international character of the process behind environmental change. Ceylon got its cocoa from Venezuela, coffee from India, rubber from Brazil. Java acquired cocoa, coffee and rubber from Ceylon. Cuba took sugar from Santo Domingo, cotton from North America and coffee from French Guiana. Island economies, although in no sense 'typical' of the kinds of transfer of resources which were transforming global agriculture, present a particularly vivid example of this process.

Plantation and settler societies under European colonialism

The environments of developing countries bear the imprint of colonial history, but the social formation established under colonial rule varied widely. In plantation societies, as we have seen, foreign investment was confined to trade, plantation crops and minerals. Plantation societies such as Ceylon, Malaysia and the island economies of the West Indies were severely constrained in what they could produce for their domestic markets, and the patterns of land use established by the colonial powers effectively tied natural resources to export production. The economic relations between the colonial power and its dependents were complementary but clearly subordinate. No competition was allowed with the colonial power. As de Silva (1982, 27) puts it: 'The infrastructure facilities (in India), geared overwhelmingly to the needs of the export sector not merely failed to benefit village interests, but involved a repudiation of these interests.'

The plantation economies were drawn into the world economy as appendages of the colonial power rather than as partners. As colonial states exerted pressures to exclude competition and secure markets and sources of raw materials, they ensured that within plantation economies the peasant sector would survive to service the plantation. Technological processes over time reinforced this domination, cementing links with the colonial power at the cost of weakening links between the peasant farmer and the land. State intervention in non-settler colonies was limited to guaranteeing captive markets for the colonial power and developing the rudimentary infrastructure required by dependent export economies. Merchant capital, although rarely owning plantation land, exercised the dominant role in the colonial export economy, conditioning the pattern of investment and surplus utilization (de Silva 1982, 37).

The bias of plantation agriculture was anti-technological, and the plantation labour process was largely labour intensive. Through commercial houses and colonial agents, such as the British Crown Agents, a measure of judicial and technical control was exercised over the way natural resources and labour were employed. Struggles to acquire land in such societies have a strong class character, since plantation workers are not peasants and produce few food staples. The vulnerability of the plantation to external markets, together with the penetration of merchant capital, have made diversification of production difficult. The horizontal organization of production has served to open plantation systems to agribusiness on a world scale, linking corporate control of marketing and processing to the local ownership and control of natural resources.

Settler societies developed differently. Countries such as South Africa, Kenya, Rhodesia and Algeria all contained large European populations who adopted colonial status. The native population was usually sparsely distributed and the relatively numerous Europeans sought a permanent presence on the land. The growth of urban centres and the diversification of economic activity led to increased tension with the colonial power. Manufacturing output grew and agriculture was concentrated not simply in export products but in grains, livestock and fruit farming. Development became internally oriented, as the large volume of production for total and regional markets led to relatively high wages and interest rates. This contributed to improvements in the application of technology, especially to agriculture. In settler societies divisions either occurred between the resettled native population and the whites (Mau Mau in Kenya) or took the form of counter-insurgency (as in the Portuguese colonies and South Africa). An alternative was to challenge the colonial power's hegemony directly (as in Algeria and Vietnam). Countries without significant indigenous populations (Australia, New Zealand, Argentina) quickly moved to privatize the land resources of the native peoples, frequently pushing them into starvation.

It is important not to lose sight of the fact that, by the end of the nineteenth century, many of the crops which had been cultivated for the colonial powers had also been transferred to peasant producers. Cocoa, which had been established on plantations in Ceylon, Java and Brazil, became a peasant crop in West Africa. Rubber, which had been developed in Brazil in the eighteenth century, was transferred to Asia in the early twentieth century. The stimulus provided by motor transport in the developed countries led to rubber plantations under the control of small farmers in parts of Malaysia, India and Ceylon. Until the nineteenth century, as we have seen, most of the crops for the European markets were condiments and quasi-drugs (spices, sugar, tea, coffee, tobacco). The demand for all these products rose with the rapid urban growth of population in Europe and North America in the twentieth century. New products also grew in importance: rubber, sisal and cotton were important industrial raw materials. At the same time food staples, particularly grain and meat, were grown for the European market from the late nineteenth century onwards, and transported in refrigerated ships from the United States, Canada, Australia and New Zealand.

The global development of agriculture has had important ecologial consequences, the importance of which has only been grasped recently. First, the sensitivity of plantation monocultures to disease led to extensive use of chemical methods to control pests and plant diseases. Insecticides and pesticides have often proved highly effective in reducing water-borne diseases, such as the eradication of malarial tsetse fly. In many cases, however, heavy reliance on the use of pesticides has proved an obstacle to natural systems of pest control, and pests have developed immunity to chemicals (Weir and Schapiro 1981, Bull 1982). Pesticides can also prove a danger to the labour force which uses them; according to the World Health Organisation, someone in the underdeveloped countries dies from pesticide poisoning *every minute* (Weir and Shapiro 1981, 3).

The internationalization of agriculture also brought major disruptions in ecological cycles, as crops were no longer produced and eaten in the same place. Traditionally crop production had led to the disposal of wastes to the ground, maintaining the organic content and trace minerals in the soil. Today the international character of agricultural production and distribution means that surpluses are transported thousands of miles. Organic wastes, instead of being returned to the soil, are dumped in rivers and oceans. Industrial fertilizers, as we saw in chapter 2, are routinely used to offset this decline in natural fertility. However, their use is expensive in financial terms and can also have harmful ecological effects by increasing the nitrate levels in water sources and rivers.

Finally, the development of global agriculture has led to increasing genetic manipulation of plant and animal species, particularly by corporations and international seed-banking companies. In their natural state the availability of genetic materials has been critically affected by tropical deforestation and other activities linked to agro-industrial development. As we shall see in chapter 8 the latest phase of agro-industrial research, in biotechnology and genetic engineering, transfers natural species from the wild to the laboratory, from the food chain to the kitchen. The discovery of new ways of reproducing nature thus promises possibilities of reducing our dependence on an increasingly eroded and polluted biosphere.

Agro-industrial expansion and the international division of labour

The increasingly important role of transnational corporations in agriculture and the food industry was discussed earlier. It was suggested that transnational corporations (TNCs) are playing an enlarged role in developing countries partly because of the logic of their own development. Technological changes, in food-processing and distribution, as well as farming, have served to integrate different stages of the food system which are located in different countries. The provision of inputs to farming in low-income countries needs to be considered together with market demand in the North. In assessing the impact of agro-industry, particularly that part of it controlled by TNCs, we need to consider several related issues, including the environmental effects of changing production technologies and the nutritional effects of changes in the diet.

Agro-industry is the name given to an integrated system comprising the production, processing, marketing and distribution of food and fibre products, in which farming becomes part of a vertically organized, industrialized process. It is important to emphasize that not all agro-industry is transnational and the TNCs are not exclusively involved with the food industry. Nevertheless, there are several reasons why TNCs have taken the lead in agro-industrial expansion. First, the activities of transnationals are inevitably bound up with the question of food surpluses and deficits in the developed world. As we have seen, overproduction in the North has led to periodic dumping in the South. Second, TNCs need raw materials and labour that can be provided cheaply only by developing countries. Third, TNCs seek to operate internationally to make, and protect, their profits. Even financial losses in the South can be preferable to them, for fiscal reasons, to concentrating their activities exclusively in developed countries.

Important shifts are currently occurring in the international divison of labour which governs which parts of the world produce which products. As intensive livestock production assumes more importance in the North, the developing countries have shifted away from dependence on plantation crops alone and now supply animal feeds to an increasing extent. Soya from Brazil and manioc from Thailand feeds livestock in Western Europe both for domestic consumption and (in the case of broiler chickens, for example) for re-export to some developing countries.

The operation of TNCs thus radiates decisions which transcend their immediate environment, relocating the production of crops and the rearing of animals in ways which meet the needs of agro-industry. Many of these effects also transcend considerations of legal ownership, helping to bring about changes in land tenure and migration patterns. As the food system widens, so more activities fall within the net or become marginalized from it. Agro-industrial expansion also modifies the conditions of capital accumulation in the remaining links of the production chain affecting, for example, the transition from subsistence to commercial agriculture among producers who function as suppliers.

It is important to recognize too that TNCs based in the United States invest most of their capital in other core countries. According to the United Nations, United States agribusiness invested half its overseas capital in Western Europe, 34 per cent in Canada and only 16 per cent in developing countries (UN 1982). Western Europe and Canada, of course, have large internal markets for the products of agro-industry and considerable existing processing capacity. Most agro-industrial capital investment is in the industrialization of food and fibre, not its production. This does not prevent the finished product's being sold in developing countries, where little local competition exists, although such sales only represent a small part of total sales. Relatively few countries in the South account for the lion's share of TNC investment: according to the United Nations in

1978, 55 per cent of United States capital invested in Latin American agro-industry was confined to Mexico and Brazil, a slightly smaller proportion than total United States investment in the region (UN 1982).

The importance of agro-industrial expansion is not confined to large countries however. For small countries their *relative* share of transnational investment in agro-industry is much less important than their share of *national* investment. As a recent survey showed, 'of the smaller countries in Latin America only Bolivia and Panama do not find their most important enterprises clustering around the agro-industrial sector' (*South* 1985). Small countries in Latin America such as Costa Rica, El Salvador and Paraguay, are among those with the largest TNC penetration. In El Salvador, for example, five of the leading private companies are involved in agro-industry, and their turnover dominates the corporate sector of that country.

As we have seen, the increasing technical sophistication of large-scale agriculture in developing countries is closely linked with agro-industrial systems, and heavily dependent on state support in Africa and Latin America. Governments in both these continents subsidize both the inputs to agro-industry and (through guaranteed prices) the principal users of these inputs, the large farmers. Since the most profitable area is that of 'value added' at the processing stage, it is not surprising that governments have concentrated their help at this point. It also helps to explain an apparent paradox – that agribusiness often favours collaboration with associate producers rather than direct ownership of land. This kind of association, which sometimes takes the form of contract farming, enables TNCs to reduce their production risks and avoid political problems. Most governments in developing countries which seek to defuse land conflicts and acquire new agricultural technology do so by approaching foreign agribusiness or consulting agencies. It makes little difference whether agrarian reforms have been undertaken: two Latin American countries in which TNC penetration is greatest, Mexico and Brazil, are quite different in this respect, but the importance of TNC involvement in agro-industry is similar in both cases.

The importance of considering the international dimension of food systems is easily recognized if we change our focus and concentrate on the logic of TNC expansion for core areas like North America and Western Europe. The United States is also the world's largest agricultural exporter. In 1980 one in every three acres planted in the United States was dedicated to export products, valued at US$40 billion (Luiselli 1985, 36). There is an imbalance of trade between the United States and most other countries, in favour of the United States. United States companies are in a correspondingly powerful position to dictate the terms under which other countries produce food for the American market. The agricultural sector in the United States itself is a smaller part of the gross domestic product than formerly. It is also a bigger net consumer of inputs produced in other sectors: energy and transport for example. Farms in the United States have tended to become individually more specialized, as they have become more diversified nationally. The United States agricultural sector has become more integrated nationally, while concentration within it has increased (Luiselli 1985). The implication for developing countries is clear: the impact of the American food system on poor countries is an elaboration of this process of specialization/diversification. TNCs, most of which are American owned, are strategically important to American agriculture by providing markets, raw material supplies and inputs which are not provided for by the national food system. For example, United States companies seeking to make up a shortfall in water supplies, or in energy, find suppliers south of the border. The Mexicali valley in northern Mexico not only supplies water to prosperous Californian towns, it also provides fruit and vegetables to the American market. As Whiteford observes, 'in each case the changes [in the valley] resulted from developments outside the region' (Whiteford 1986). These patterns of resource use, extending outside the boundaries of the United States, are complemented by migration of Mexican labour to the United States (Kearney 1986, Wiest 1984).

Another important aspect to consider is the demand for food within the United States, which has an important, albeit indirect, effect on the environment of countries to the south. Changes in the diet of most Americans have been both cause and effect of changes in the American food system. The food system has become more capitalized and costs have been absorbed by technical advances. Changes in American eating habits have been transferred to developing countries, through media exposure and the establishment of food-processing plant for the internal market in other countries. United States companies have sought to plug the gaps in their supplies by importing beef from Brazil and Central America (Hecht

1985) and fruit and vegetables from Mexico (Rama and Vigorito 1979). In their way these strategies parallel those of water policy in the Mexicali valley – they make sense in terms of the requirements of American consumers, rather than of the particular advantages of agricultural production and land use in Latin America. However, the imbalance in agricultural trade between the United States and Latin America means that even quite small shifts in American domestic demand bring large shifts in the use of land, technology and energy in Latin America.

As changes have occurred in food consumption in the American market, so the internal market of Latin American countries has accommodated to and reflected these changes. Most Latin American countries have a heavily skewed distribution of income and wealth among the population. The state has proved important in such cases in stimulating agro-industry to meet the needs of the domestic middle class. Some writers even argue that the indirect effect of changes in the internal market, partly induced by agro-industrial expansion, is more important than the effect of trying to meet export demand: 'Internationalized agribusiness molds and covers the food needs of the middle class, which in turn provides economic and political support for this modern, productive force' (Spalding 1984, 14).

As we shall see later in this chapter, the development styles adopted by most Latin American countries have been designed to meet the needs of their middle classes, rather than help secure sustainable growth and development. In most so-called 'Newly Industrializing Countries', in Asia as well as Latin America, the domestic middle class is an important basis for the international restructuring in the world economy which is taking place in the 1980s. In the Latin American case social classes which grew to prominence through state-supported industrialization in the 1950s and 1960s have proved important elements in ensuring that agro-industry continues to receive support, both from national governments and from international agencies.

One effect of the continuation of domestic 'adjustment' policies, aimed at the public sector, and the expansion of agro-industry in large-scale privately owned agriculture has been to increase the gulf between peasant rain-fed agriculture in Latin America and the relatively prosperous irrigated zones. As Sanderson puts it for Mexico:

as stabilization proceeds, the mandate for reinstating real price controls in the current climate of inflation will further discourage the agricultural bourgeoisie of the federal irrigation districts and better rain-fed lands from planting basic crops, instead of the uncontrolled agro-industrial inputs and cattle feed crops that many of them concentrate on now . . . the agricultural crisis of Mexico – whether considered as an employment, nutritional, distributive or legitimation crisis – is the product of the everyday mechanisms of international integration, rather than the episodic shudders of world recession or the fluctuations of oil prices (Sanderson 1986, 274).

International food policy and the environment

Changes in the international division of labour are not only promoted by transnational corporations, they have also been fostered by the national and supra-national food policies of the industrialized countries. An example with which we are particularly familiar is that of the European Economic Community (EEC) under its Common Agricultural Policy (CAP). The CAP was established to help farmers within the EEC benefit from the increased application of technology to agriculture. It has – almost incidentally – assumed importance as a mechanism through which European consumers have denied developing countries access to their domestic markets on competitive terms. In underwriting the European farmer the CAP and its accompanying regulations have placed a considerable additional burden on subsistence farmers in the poor countries.

The CAP has functioned as an instrument to help create a low-risk economic environment in which farmers were able to specialize in agricultural production, abandoning the mixed-farming tradition of their forebears. It has also served to accelerate regional agricultural specialization within the EEC. Prices have held stable above world levels and, in addition, farm produce from the EEC has been offered a preference in the domestic market, while receiving export refunds on the international market (Bowler 1985, 134–5). This specialization has tended to increase the structural rigidity from which European farming has suffered as farmers became more dependent on fewer products. In addition it has increased the farmers' dependence on the lobbying and political processes which have succeeded in guaranteeing them secure incomes. From a historical standpoint it can be argued that sectoral- and commodity-producer groups within the EEC now constitute a powerful force for resistance to change in world trading patterns, at least equal to the kind of political protectionism which has marked foreign policy in Europe for much of this century. They have also played a part in generating agricultural surpluses and budget crises which have led to the growing conviction that current allocations of funds are themselves unsustainable.

The principal effect of European farm protectionism has been to create surpluses which need to be stored or disposed of. Until the imposition of milk quotas on EEC farmers, the growth in the dairy surpluses was unbridled. After 1976 the scale of these dairy surpluses was contained within more manageable limits. Between 1976 and 1980 (milk quotas were first imposed in 1977) the total of 1,135,000 tonnes of dried milk in store within the EEC was reduced to 180,000 tonnes, but this achievement was only at an enormous budgetary cost to EEC taxpayers and consumers.

A similar situation exists today with grain surpluses. Within the EEC cereal production has risen rapidly, almost entirely as a result of improvements in yield. Since 1970, for example, average wheat yields in the EEC have increased from under 4 tonnes per hectare to over 6 tonnes per hectare (Hubbard 1986, 197). The area sown to all cereals within the EEC has also remained much the same. The result is that, without corrective policies to reduce cereal production, surpluses would be expected to reach around 40 million tonnes (over a third of cereal consumption) by 1991/2. The management of this 'grain mountain' is expensive, since the EEC pays its cereal farmers much more than the price it receives when exporting the surplus outside Europe. This difference is paid by the taxpayer. In addition, storage costs are high, and it is difficult to dispose of grain surpluses on the international market even at subsidized prices. Competition exists from the United States, Canada, Australia and Argentina, all of which are attempting to export their surpluses. The cost of the cereal sector to the countries of the EEC is enormous: within the overall budget of the CAP it is currently around £1.5 billion a year.

Subsidies of this kind not only pre-empt EEC resources which could be used for other purposes, but they also shut out poorer countries from selling their grain to the EEC. The absurdity of the situation can be grasped if we substituted ships for grain under the kind of arrangements employed for the CAP. This would mean that shipbuilders could build as many ships as they wanted, protected by a 100-per-cent tariff wall from the outside world, with the government freeing them from the need to pay rates, and guaranteeing to buy every ship they could produce at prices well above world market rates (*Guardian*, 6 June 1986). What would clearly be an intolerable degree of government support in shipbuilding is, in fact, public policy in agriculture – not merely in the EEC but in other developed countries like Japan and the United States.

Table 5.1 Estimates of land displaced in developing countries by imports of forage crops into European Economic Community

Account of non-Community-produced fodder in livestock production

	(%)	Equivalent area (production using non-Community fodder divided by national or Community cereals yield per hectare)
FRG	23	3,000,000 ha
France	9	1,200,000 ha
Italy	26	2,200,000 ha
Netherlands	52	2,400,000 ha
Belgium/Luxembourg	38	900,000 ha
United Kingdom	17	1,300,000 ha
Ireland	10	200,000 ha
Denmark	23	700,000 ha
Community of 9	18	9,700,000 ha

Note:

The proportion of world cereal production fed to livestock increased from 37% in 1961–5 to 41% in 1975–7. In addition, industrial production of animal feedstuffs is one of the main features of agricultural development since the beginning of the 1950s: the current level of production in France is 15–16 million tonnes per year, or one-third of the total cereal crop. These figures constitute a problem in relation to hunger in the world, especially since Europe imports, on average, 42.4m. tonnes p.a. (1980–2) of animal fodder. To some extent, therefore, we use land in the Third World to feed our livestock, mainly at the expense of subsistence crops. For example, soya cultivation has spread in Brazil at the expense of black beans (see Y. Chavagne, *L'agriculture industrielle en crise* (Industrial Agriculture in Crisis) (1984) Paris, Syros; B. Carton, 'Frères des hommes').

Source: Thiede 1986, 18

The negative effect of EEC policies is compounded by the preferential treatment given to those agricultural exports from developing countries which can be used to feed animals within the EEC (see table 5.1 above).

The rather absurd situation exists whereby livestock production in the EEC has become increasingly dependent on imports of cereal substitutes (protein) from developing countries (6.35 million tonnes in 1979) while the export of cereals is subsidized (Bowler 1985, 89).

The EEC is now the world's largest importer of animal feed products, specializing in the production of high-value foodstuffs from cheap components provided by developing countries. Products such as soya are imported from Brazil and manioc from Thailand to be fed to livestock in intensive feed lots within the EEC. The impact on the farming systems of the supplier countries is necessarily considerable, as resources are dedicated to converting land to export production, often from subsistence or food-crop-producing systems. In some cases the animals reared in Europe are 're-exported' to developing countries in which the skewed distribution of income has created a domestic market – broiler chicken exports to the Middle East are a case in point. The result again is that international trade in agricultural products is orientated towards meeting the needs of rich consumers and undermining those of poor farmers.

There is real substance to the charge that the conversion of vegetable to animal protein within the EEC has been a means of effecting trade relations that are injurious to developing countries. For example, the export of beef from the EEC was only 5 per cent of world trade in 1977, but 21 per cent by 1980 when EEC exports began to displace beef exports from Argentina to Egypt, and from Uruguay to Ghana. The effect of EEC trade policy in foodstuffs has thus been to damage relations between countries in the South, as well as between the South and the North.

The response of the EEC to its transformative role in international agriculture is to emphasize the positive aspects of policies designed to help developing countries, and to stress the efforts that are currently being made to reform the provisions of the CAP. The EEC sees the Rome treaty arrangements with African countries as part of its supra-national development effort. Nevertheless this needs to be balanced against the negative effects of the EEC's 'food aid' policy, which has tended to be used as a way of dumping unwanted surpluses rather than developing African agriculture. The indirect effects of EEC food aid include the undermining of small food crop producers in some African countries. By 1982 the subsidy to agricultural exports in the EEC had reached 35 per cent of the Agriculture Fund – the remaining money being used to support agricultural products on the internal market. Together these represented the 'Janus face' of the EEC: subsidizing uncompetitive producers and the export of their surpluses.

Attempts to reform policies within the EEC have failed to grapple with the underlying structural inequalities of present policy. For example, reform of cereal production is to be undertaken through a so-called 'co-responsibility levy' which would seek to claw back from farmers some of the price advantages they received for their cereals. This would then be used to help finance the cost of surplus disposal. Within the EEC the farming lobby probably regards the co-responsibility levy as preferable to any of the other options, which might include a quota on the amount of grain produced (like the milk quotas) or the reduction of prices to the producer. However, the co-responsibility levy is a purely financial solution to the problem of overproduction, which does little for the EEC taxpayer and less for the developing-country farmer. As Hubbard has argued, 'the levy has the attraction of giving a general impression that the Community's farmers are helping foot the bill for excess production' (Hubbard 1986, 199).

The reality is probably different. It is likely that European farmers will be relieved of the burden of the levy by a corresponding increase in the support price. Then the consumers within the EEC will find themselves supporting farmers via slightly different channels: the net producer price plus the co-responsibility levy. The central point is that a co-responsibility levy for cereals is designed to raise revenue to help cover the cost of two unnecessary processes – the cost of storing grain surpluses and the subsidy which will continue to be devoted to the export of these surpluses. Whatever the appearance of co-responsibility policies – and European farm groups will be quick to exploit their liberal image – they in no way alter the structural policies which were erected to defend European farming interests against those of developing countries.

Finally, the environmental effects of developed-country food policies are not confined to the transformation of land use in the poorer developing countries. Within the developed countries intensive animal production raises other environmental problems, such as those of animal welfare and the disposal of slurry through rivers and watercourses. The increasing specialization of production has also served to change the landscape in many ways and has contributed towards the reduction of natural habitats for birds and wild animal species. Dependence on heavy machinery has increased soil compaction, while the nitrification of water resources has been one consequence of washing fertilizers from the land into drainage systems. Changes in the environment, themselves the consequence of international restructuring and the development of agro-industry, are in no way confined to the South. It needs to be re-emphasized that the internationalization of the environment is a process which transforms both developed and developing countries, often in ways which initially appear unconnected and separable, but which form part of wider, systemic changes in the relationship between land uses and consumer markets. It is worth examining these relationships in a specific region of the South, taking Latin America as a case study.

Development styles and an alternative development model: the Latin American case

The effect of changes in the international division of labour on the environment has recently been analysed in a series of excellent studies undertaken by the joint Development and the Environment Unit of ECLA/UNEP in Santiago, Chile, since 1980. During the execution of an initial project on Latin American development styles between 1978 and 1980, it became clear 'that it was necessary to deepen understanding of the relationship between the environment and the characteristics of development in Latin America' (CEPAL 1985a). It was concluded that in order to assess the viability of current development in the region it was 'not enough to maintain and expand the social capital endowment in its traditional categories, but there must also be an explicit recognition of the close interdependence between society and nature' (CEPAL 1985a). This recognition has led to a corpus of well-researched, systematic studies of how economic growth and development might be reconciled with sustainability over the long term and in specific local environments. These studies and the analysis to which they gave rise have assumed increased importance as the international repercussions of the debt crisis and economic restructuring have increased in Latin America, as in other less developed areas. The best examples of this work include specific studies which compare the effects of domestic economic policy and international restructuring on specific ecological regions and labour processes.

Capitalist countries are usually classified into two groups: those which are central to the global economic system and relatively developed, and those which are peripheral and poor (Prebisch 1976). This distinction is of importance in characterizing the living conditions of people in a continent like Latin America, where even the most developed countries like Mexico and Brazil are still peripheral to capitalism. As Lefever and North have pointed out (1980, 5):

> there is a fundamental difference between the social welfare effects of capitalism in the industrialized nations and the process of development currently taking place in Latin America. In contrast to the earlier history of the industrialized countries, which in its own tortuous way did lead to significant increases in working class living standards, capitalism in Latin America has failed to spread the benefits of economic growth to the lower income groups.

The feature of Latin American societies which has received attention from most authors is the high degree of social differentiation between the élite groups and the great mass of the poor. Prebisch highlighted this process in the following way:

> peripheral capitalism, particularly in Latin America, is characterised by a dynamic which excluded the great mass of people. It is a dynamic process orientated toward the privileged consumer society. This is because the process of capital accumulation and the introduction of new technologies from industrial centres are not motivated by the purpose of progressively incorporating new social strata in the development process (1976, 21).

Expressed differently, Prebisch is saying that the industrialized countries need to incorporate their own workers as consumers of their manufactured goods, so the benefits of development reach throughout the economy and bolster the internal market. This internal market is the means of creating and maintaining steady industrial development. The situation is different in peripheral countries which mainly produce raw materials and primary industrial goods and whose populations do not provide as much consumer demand for these goods. De Janvry, for example, has looked at the depressed condition of the 'peasantry' in Latin America in just these terms (1981). What he terms 'functional dualism' is, in essence, a system which makes use of the cheap labour power provided by peasant families whose consumption needs are partly met within the household economy. In strictly 'economic terms' such families do not 'need' to earn more than the minimum or to participate as full members of industrial consumer society. The dualist structures in which they function require their labour power but not their consumer power. The primary influence on the development of these structures is the pattern of industrial capital ownership and demand located in the central economies. Reversing these domestic structures has proved difficult in Latin America, as the experience of Chile's period of 'substitutive' industrial development proved between the 1930s and the military coup of 1973 (Rivera 1985, 16).

The dominance of primary goods in the export structure of areas like Latin America is marked, despite changes in the composition of those goods. For example, in 1975 some 86.4 per cent of Latin America's exports were primary goods, mostly agricultural. In addition the deteriorating terms of trade for these goods on the world market has meant that countries in Latin America have had to increase their exports in order to maintain the same level of income. As a result Latin American 'agricultural production for the internal market has been neglected and only agricultural exports have grown' (Lefever and North 1980, 8). In many cases the per capita production of food has declined, as low productivity of food crops, combined with population increases, has reduced the self-sufficiency of Latin American countries. For peasant producers, as well as for commercial agricultural enterprises, supplying the internal market does not offer an acceptable level of profitability. The expansion of commercial crops, such as sorghum for animal feeding in Mexico, or soya for export in Brazil, has tended to occupy family producers as well as capitalist farmers. Most dramatically of all, good quality land is devoted either to the production of energy substitutes to 'feed' industry, such as sugar cane in Brazil, or cattle are raised under 'frontier' conditions to supply the demands of the external, primarily North American, market.

Two main trends can be identified in Latin America: one promoting industrialization for the internal market, and the other trying to achieve development through the export of primary goods. These trends constitute what some Latin American writers call 'styles of development'. The concept 'style of development' captures a number of closely related variables: economic thought and policy (such as 'neo-liberalism' in Chile), the combination of available capital and human resources, and the classes which benefit from a particular pattern of capital accumulation. From the beginning, the concept was used to characterize 'a specific combination of resources in order to reach an expected goal' (Rivera 1985, 18). According to most writers, style of development is quite a flexible concept. In general it seems to suggest a long-term attempt to build a particular social formation supported by specific social classes, given available natural and human resources. Style of development implies a class project backed by a specific economic model, which may emphasize the economic or the social participatory dimensions of development (Rivera 1985, 20). According to Rivera we can identify several such styles in Latin America's development during the last half century.

The 'industrializing style of development' was initiated after the 1929 world crisis. Its main objective was that of creating industrial development through the establishment of basic strategically important industries, directed mostly towards the internal market. During the 1950s and 1960s, partly as a response to growing awareness of the deficiencies of Latin America's agricultural sector, highlighted in the studies of the Inter American Committee for Agricultural Development (CIDA), more emphasis was given to promoting the agrarian sector. Ostensibly this took the form of promoting agrarian reform throughout the region, but in practice what proved much more important was the modernization of the *hacienda* system that this reform process largely concealed (Goodman and Redclift 1981). Between 1964 and 1980 these modernizing tendencies took different forms: for example, in Chile towards a socialist system under Unidad Popular and 'neo-liberalism' thereafter; in Peru nationalist reform under the military government that took power in 1968; and in Brazil where the military government stimulated the industrial penetration of agriculture after 1964. By the end of the 1970s Latin American agriculture had been subjected to considerable change, bringing a transformation in the rural labour process and a capitalization of large-scale agricultural enterprises. The worsening debt situation after 1982 exacerbated the effects of the restructuring that had taken place in Latin American agriculture during the previous two decades. The 'peasant economy' of the 1960s had largely disappeared, and in its place stood a depressed rural economy in which household production survived, but the dependence on wages earned in the commercial agricultural and urban sectors had increased enormously. The 'agro-food' systems that had come to dominate the commercial sector of agriculture had, to a significant degree, also penetrated the local systems of production and exchange under household management. The challenge faced by Latin American agriculture was how to generate more foreign exchange to service the foreign debt without further impairing food production for the domestic population. There appear to be two ways out of this dilemma, each of which implies important costs and benefits. They serve to illustrate the extent to which environmental issues in the South cannot be confined to questions of environmental management alone, but necessarily imply choices between development styles, strategies for reconciling limited economic growth with sustainable development objectives.

The development style which is currently being recommended to Latin American countries by the IMF seeks to reduce public expenditure and increase exports to other parts of the world on the 'East Asian' model, followed by countries like Taiwan, South Korea and Hong Kong. Latin American economies are to be further opened up to foreign capital, and the tariff barriers to external penetration removed. Further specialization within the international economic system is viewed as a necessary means of countering protectionism at home and producing greater economic integration abroad. This requires some fundamental changes in the economies of those countries, such as Mexico, which have endeavoured to support domestic industrialization in the interests of nationalism and greater economic self-determination.

The environmental critique of this development style takes issue with several of the model's assumptions (CEPAL 1985a and b). It is argued that the way in which natural resources have been utilized under the prevailing development style partly accounts for the external vulnerability of Latin American economies, since the potential of local resources has not usually been mobilized for the satisfaction of the needs of the population. Trends emanating from the fully industrialized economies have filtered into Latin American economies in a non-selective and passive fashion. This permeability has hampered the development of autonomous and flexible systems of production which could maintain their position in local and international markets. According to the UNEP/ECLA team in Santiago an alternative development path needs to be taken which meets different criteria. The following are the components of this alternative development path:

1 Proposals for utilizing natural resources should seek to generate employment and require few imported inputs.
2 Priority should be given to future resource *stocks* over present, short-term, resource *flows*.
3 Selective restrictions need to be placed on the production and importation of unnecessary goods, and a stimulus needs to be given to the supply of basic-needs goods, such as housing, food and clothing.
4 Measures need to be taken to increase the labour expended in rural conservation, for example, reforestation, terracing in areas of soil erosion, the renovation of irrigation systems, greater use of alternative energy sources.
5 In the face of declining public-sector budgets and services (itself partly an effect of depressed demand) new areas of 'collective consumption' need to be opened up and social infrastructure developed.
6 A concern for better environmental management needs to play an essential part in the training of professional groups, especially those working in agriculture (CEPAL 1985b).

What is being proposed is a sustainable-development model as an alternative way of utilizing resources, without subjecting Latin American societies to the kind of dependency which has traditionally moulded the historical development of their environments (Vitale 1983).

The search for an alternative model should not blind us to the difficulties such an approach would meet in practice. First, it would prove extremely difficult to finance environmentally sound development without external assistance. This is very unlikely to be forthcoming. Second, Latin American countries are no longer characterized by 'élites' and 'masses', as they were depicted in the 1960s (Lipset and Solari 1967; Veliz 1967). Today most Latin American countries have large middle classes, urbanized and orientated to private consumer goods rather than publicly owned natural resources. Since 1970 these classes may have provided a check on the ambitions of military groups and a basis for 'democratization'; but they are also likely to prove difficult to enlist in the cause of agricultural-resource conservation and modest economic growth. Finally, critics of the UNEP/ECLA position might brand it as hopelessly utopian. They might argue that the environment in many parts of Latin America has already been transformed in ways that are irreversible, whether in terms of food systems, the utilization of technology or the distribution of population between rural and urban areas. A sustainable future in Latin American countries, where tastes and consumption patterns have followed the fashions set in the developed countries, will require political conditions to change, and the environmental demands contained in the UNEP/ECLA position will need to become part of the political programme of social groups. These issues are taken up in more detail in chapter 7.

Managing the international environment

This chapter has argued that the environment of developing countries as well as that of developed countries needs to be considered from within an international perspective. The development of the rural resources which make up the 'environment' cannot be separated from the historical processes which link the industrialized North with the developing South. Both are bound together by economic and political ties which do not seem immediately 'environmental': trading relations, the transfer of technology, even the relocation of labour to new productive activities. Much of the 'natural' environment of developing countries is a reflection of these processes; indeed it is no more 'natural' than the rural environment of East Anglia or the Po Valley.

Public concern with international environmental problems has largely focused on issues of species survival and conservation, rather than sustainability. This is hardly surprising, given the failure of most international development agencies to grapple effectively with the problem of human livelihoods. Public opinion has been directed towards preserving non-human species of animals and plants in their natural habitats. It has been awakened, quite properly, to the damage which human actions have inflicted on nature. What it has not been awakened to is the damage which international development processes have inflicted on the environment from which human beings depend for their livelihood. In most developed countries over a third of the public is concerned about the extinction of plant and animal species and the depletion of world forest resources (see table 5.2 below). Almost as many people whose opinions were canvassed thought that changes in the tropical environment held dangers for the climate. Clearly a link exists in people's minds between what is happening to wilderness areas and the well-being of the planet in general. However, this does not necessarily extend to the links between the conservation of biotic resources in wilderness areas, and the pursuit of sustainable development in more populated areas.

The need to re-examine the concept of conserving nature in

Table 5.2 Public opinion: concern about international environmental problems selected industrialized countries

	Year	Extinction of some plants or animal species in the world (%)	Depletion of the world forest resources (%)	Possible climate changes brought about by carbon dioxide (%)
USA	1984	n.a.	n.a.	n.a.
Japan	1981	12	28	18
Finland	1983	38	41	29
EEC				
Belgium	1982	25	32	22
Denmark	1982	41	46	38
France	1982	33	34	23
Germany	1982	38	31	34
Greece	1982	32	37	34
Ireland	1982	21	21	25
Italy	1982	35	41	37
Luxembourg	1982	50	57	28
Netherlands	1982	42	33	18
UK	1982	39	40	28
EEC Total	1982	36	36	30

Note:
n.a. = not asked
Source: OECD 1985

wilderness areas has been at the forefront of recent thinking about environmental management. Recent meetings of the World National Parks Congress have brought together people from over 68 countries to discuss the development of protected, natural areas (Ingram 1983, 7). The concept of 'biosphere reserves' has been introduced, and international action has been called for to protect areas that are threatened by ecological destruction. The potential and limitation of the 'biosphere reserve' concept are discussed in chapter 7.

One of the reasons why environmental management has failed to preserve threatened areas is that the advocates of better conservation policies have usually been a small, privileged minority. In the period between 1945 and the Stockholm Conference which established UNEP in 1972, most intergovernmental co-operation on the environment was confined to protecting specific animal species under threat, such as migrating wildfowl. After 1972 'the dominant paradigm was environmental protection through governmental regulation guided . . . by an élite that had attained environmental enlightenment and salvation' (Perry 1986, 12). Although the need for intergovernmental action is tacitly recognized, the agencies which might accomplish it tend to be forums for discussion between national government bureaucracies, each with different degrees of influence. International agencies themselves tend to prefer not to encroach on the national sovereignty of member governments. Effective international action is confined to individual countries where environmental agencies receive political support from government, and there are few such cases (Perry 1986).

There are a number of international organizations with a specifically 'environmental' brief. These include IUCN (International Union for the Conservation of Nature and Natural Resources) founded in 1948, the World Wildlife Fund formed in 1961, organizations of the Council of Europe with responsibilities towards the environment and UNEP (United Nations Environment Programme) formed after the Stockholm Conference in 1972. Most international organizations have a research or, at best, an advisory role, in international environmental policy. UNEP, for example, has the role of environmental monitoring, environmental planning and the dissemination of information. Its role has always been widely misunderstood, however. It is not a UN *executive* agency, empowered to carry out its own programmes in the member states (like FAO or UNESCO). Nor is it a sprawling UN organization with a huge professional staff and a correspondingly large budget. Most importantly, UNEP is not responsible for most of the world's environment which lies within the boundaries of sovereign nations unprepared to tolerate interference from UN bodies in their internal affairs (Sandbrook 1982). The role of UNEP is to be a catalyst within the corridors of international opinion, raising consciousness of environmental issues especially within the UN 'system'.

The reality is that UNEP has little money and few staff, even fewer incentives to offer and no means of enforcing its wishes. It is a little like creating the United Kingdom's Natural Environment Research Council (NERC) without pooling the resources of its constituent member organizations and without co-ordinating their management structure (Sandbrook 1982, 333). Even the official document which was produced as Britain's contribution to the debate initiated by the World Conservation Strategy had this to say about UNEP:

At the end of the day, the unavoidable truth is that the combined resources of the Governments who came together at Stockholm to create the UN Environment Programme have simply not:

(a) funded the Programme adequately;
(b) cooperated with it adequately;
(c) intervened with sufficient vigour to improve its performance;
(d) taken much notice of it (as Governments) except when it suited their short-term ends (Sandbrook 1982, 335).

There are a number of reasons why the international environment cannot be 'managed', even when it is recognized that the nature of environmental problems makes international action necessary. As we have seen, the instruments for concerted international action are in their infancy and depend for their effectiveness on domestic political support in developing countries. The structural position of most of these countries, their poverty and indebtedness, make it extremely difficult to mount concerted action on behalf of the environment, notably when this implies (as it invariably does) external involvement in domestic affairs. Designations such as 'biosphere reserves' are important as indications of what might be achieved by concerted planning, but they are indicative rather than realizable achievements. In addition, as we shall see in chapter 7, the techniques of environmental management which are employed in most developing countries are frequently inappropriate since they fail to involve the local population sufficiently.

Finally, the management of the international environment can only be successful if it is fully recognized that environmental concerns cannot be divorced from development policies. As we shall see from the next chapter, the transformation of the environment in developing countries is linked to the development of market economies and commodity production, which have served to undermine traditional resource uses and the ecological systems on which they depended. To fully appreciate the nature of this process we need to return to historical experiences.

6

The transformation of
the environment

It is interesting to compare the following accounts of the same place, the Valley of Mexico, which describe the situation in 1984, and some 470 years earlier.

1 Valley of Mexico (1984)

The following description of the Valley of Mexico, provided by the Office for Urban Development and Ecology of the Mexican Government, formed part of its submission to the Man and the Biosphere Programme (MAB) of the United Nations (Sanchez de Carmona 1984).

The Valley of Mexico is located in the extreme south of the central *mesas* and covers a surface area of 9600 square km. The land is suitable for crops, fruits, natural pastures and man-made grasslands. The main land use problems are linked to the lack of soil nutrients, erosion, salinity, alkalinity and flooding. Overgrazing and excessive deforestation have led to extensive resource depletion in the valley.

. . . (There are also) serious pollution problems, due as much to the wastes resulting from domestic daily activities as to wastes from large industrial zones. Pollutants are emitted from industry, motor vehicles (2 million cars) and 'natural' dust storms which originate in the areas around the dried-up Lake Texcoco in the N.E. of the city and which blow human waste from these sewage outlets all over the Federal District.

Waste disposal is a major problem. The metropolitan area produces 6000 tons of solid waste each *day*, of which only 75% is collected. The rest is scattered throughout the city, most of it on open untreated dumps. Those in 'marginal' settlements near the dumping grounds are most severely affected by these wastes and there is resulting environmental degradation, through methane production, and through soil and water contamination.

The population, currently about seventeen million, is expected to reach about thirty million by the year 2000. Today water is transported long distances across mountainous regions to the Federal District. Electrical consumption for pumping water may well double between 1985–88 and then *double again* by 1990 if the waters from the Tecolutla Basin have to be raised about 2000 m over a distance of 200 km. At present 50% of the land surface is affected by problems of erosion. If deforestation, overgrazing and inappropriate agricultural practices continue as at present, an even more critical situation will result. The problem is compounded by continued urban incursion into highly productive agricultural land which results in a lowering of agricultural production.

2 Valley of Mexico (1519)

The following extracts are from Bernal Diaz's account of the same place, four and a half centuries earlier (Diaz 1963):

During the morning (8th November 1519), we arrived at a broad causeway and continued our march towards Iztapalapa, and when we saw so many cities and villages built in the water and other great towns on dry land and that straight and level causeway going towards Mexico, we were amazed and said that it was like the enchantments they tell of in the legend of Amadis, on account of the great towers and cues and buildings rising from the water, and all built of masonry. And some of our soldiers asked whether the things that we saw were not a dream . . . With such wonderful sights to gaze on we did not know what to say, or if this was real that we saw before our eyes . . . (with Montezuma), we went to the orchard and garden, which was a marvellous place both to see and work in. I was never tired of noticing the diversity of trees and the various scents given off by each, and the paths choked with roses and other flowers, and the many local fruit trees and rose bushes, and the pond of fresh water . . . We must not forget the gardens with their many varieties of flowers and sweet-scented trees planted in order, and their ponds and tanks of fresh water into which a stream flowed at one end and out of which it flowed at the other, and the baths that (Montezuma) had there, and the variety of small birds that nested in the branches, and the medicinal and useful herbs that grew there . . . I may add that on all the roads they have shelters made of reeds or straw or grass so that they can retire when they wish to do so, and purge their bowels unseen by passers by, and also in order that their excrement shall not be lost. . .

We saw the three causeways that led into Mexico . . . We saw the fresh water which came from Chapultepec to supply the city, and the bridges that were constructed at intervals on the causeways so that the water could flow in and out from one part of the lake to another. We saw a great number of canoes, some coming with provisions and others returning with cargo and merchandise; and we saw that one could not pass from one house to another of that great city and the other cities that were built on the water except over wooden drawbridges or by canoe . . .

The city which the Conquistadores discovered in the Valley of Mexico on 8 November 1519 was a remarkable island-capital which covered over 20 square miles. The combined population of Tenochtitlan and Tlatelolco was between 200,000 and 300,000 people, five times the size of Henry VIII's London. Indeed the population of Mexico as a whole was probably in the region of 11 million, many times that of England.

What surprised and fascinated Bernal Diaz most was the 'agricultural' nature of the city he discovered. It was divided, on a grid system, by long canals intersected by river 'streets'. Between these 'streets' were rectangular plots of land with houses built on them. These were the *chinampas*, the raised vegetable beds which provided most of the produce consumed in the city. These raised beds had been known to the lowland Maya during their 'Classic' period – as long ago for the Aztec population of Tenochtitlan as their civilization is to us. Constructing canals from the thick marsh vegetation, the Aztec people had piled up the surface vegetation like green 'mats'. Then, from the bottom of the canals, they had used mud to spread over the green 'rafts', which were anchored by planting willows all around. The fertile plots that were constructed in this way produced a variety of crops, vegetables and fruit trees. Houses were built of light cane and thatch and, on drier ground, even of stone and mortar.

This enterprise met with problems which had to be overcome. Communication was by way of planks laid over the canals. To reduce salinization of the water supplies, as the lake was high in salt content, a ten-mile dyke was constructed, sealing off a spring-fed freshwater lagoon for Tenochtitlan. Through human ingenuity the Aztecs were able to turn environmental obstacles to their advantage.

The chinampas were also extremely productive. As late as 1900 they still supplied some vegetables to Mexico City from the much reduced Xochimilco beds, all that remained of the chinampa capital. Three harvests were possible, with transplanting from reedbeds; animals were kept and their manure (together with that of humans) used on the organic gardens. In recent years there has been a growth of interest in raised-bed systems in Mexico and elsewhere (Toledo *et al.* 1981, Morales 1984).

Our interest in chinampas, however, need not be confined to their current agronomic potential, important as that is. The accounts of pre-Columbian sustainable agriculture should also lead us to more fundamental questions about 'development' itself, and the role of the environment in the development process. Should we dignify with the term 'development' a process which leads millions of people to sacrifice their health and energies to survival? Perhaps an ecological alternative lies not so much in learning things we do not know, as in 'unlearning' things we do know?

Within 50 years of Bernal Diaz's arrival the cities of Tenochtitlan/ Tlatelolco were pale shadows of their former selves. The pre-Columbian hydraulic system, analogous in the New World to the systems that had raised the ancient civilizations of China and the East (Wittfogel 1981), had been irrevocably destroyed. Instead the Spanish had established patterns of resource use which maximized the 'tribute' they received from the subordinate indigenous population. In place of the delicate ecosystems that had once supported millions of people, they attempted to mobilize labour for the production of agricultural surpluses. These surpluses were essential to the mining economy, and later the great estates, which formed the basis of Spanish colonial power. The first post-Conquest agricultural production crises in Mexico, especially in 1538 and 1543, were what we would term 'resource crises' today. The attempts to utilize the water from the lake complex for irrigating new land and flood prevention were largely unsuccessful. At the same time the rapid growth of urban population throughout the colonial period could not have been supported by raised-bed agriculture, which was probably more vulnerable to demographic pressure than a system based on tribute and the production of agricultural surplus for an urban population.

By the eighteenth century the establishment of the *hacienda* or large estate, together with the use of plough and animal traction, enabled the Valley of Mexico to achieve greater self-sufficiency in agricultural production. However, most of the indigenous population was deprived of land and pushed into increasingly marginal areas where subsistence production was practised. The colonial system was capable of higher levels of accumulation than the Aztec empire – but only at the cost of producing two social formations held together by dominant, colonial rule. The need to clear much of the land of forest, together with the insatiable demand for wood from Mexico City (which required 25,000 new trees a year during the late colonial period), meant that deforestation was severe in the valley. Both indigenous communities and haciendas played a part in this decline (Gibson 1981, 312 and 362–4).

Changes in land use have been even more dramatic during this century. The hacienda system established under colonial rule was largely replaced in the post-Revolutionary period by a combination of *ejidos* and peasant communities (PRUSDA 1984). In a sense this marked a return to greater self-provisioning in agriculture. However, the urban growth of Mexico City dictated a quite different pattern of resource use from that established in pre-Columbian times. Food supplies for the urban economy came from further afield and the immediate environment became important for the provision of two other commodities: land and labour. The need for cash income, with which to enter into exchange relations, ensured that peasant agriculture in the Valley of Mexico existed in name only. Most of the population in the peri-urban area and much of the land became incorporated in the increasingly centralized and specialized growth of Mexico City. In the space of four centuries an ecological system built on sustainable agriculture (chinampas and terrace cultivation) had been replaced by one in which labour and land become first separated and then recombined (the hacienda/community) under a new technology (plough and animal traction).

The objective, to ensure control over the economic surplus, was achieved at the cost of destroying the indigenous ecological and cultural systems. Finally, the redistributive changes ushered in by the Mexican Revolution of 1910 provided a means to 'modernization', the ultimate effect of which was to increase social inequality and place the environment in jeopardy. Mexico City became the material representation of a new kind of development, in which in the short term economic growth and the carrying capacity of the ecological system increased, but which ultimately undermined the long-term stability of the resource base on which the rural population depended.

The dramatic transformation of the environment in the Valley of Mexico over a period of almost half a millennium illustrates what can happen when a colonizing power imposes solutions which are inappropriate to the environmental conditions they meet, however closely they reflect the conditions in the society from which they came. Colonialism, in Africa and Asia as well as in Latin America, has frequently distanced itself from sustainable solutions, precisely because the principal objective is no longer subsistence, but the concentration of power and the accumulation of capital. It also suggests that changes in the relations of production, which govern the new technologies, cannot be understood without considering the wider systems of exchange and specialization which are essential to urbanized, industrialized society. This chapter discusses the process through which the environment is transformed from a local cultural and ecological system, usually with external linkages, into a functioning element in the international system.

Chapter 5 looked at the internationalization of the environment from the perspective of historical changes in the international economy and the current international division of labour. This chapter takes local environments and societies as its focus. It becomes clear that, viewed from the perspective of small-scale societies, the development of capitalism does not imply a sequential, linear transition from one kind of environment to another but, rather, important shifts in the relationship between people and their environment. In the course of the transition to capitalism, which Eric Wolf reminds us has had a very different impact in the various phases of development (Wolf 1982, 305) it is not simply the environment that changes but peoples' perception and knowledge of it. The essentially dialectical relationship between people and their environment has no predetermined outcome. Human consciousness of environmental issues is a product of changing material conditions, themselves forged from human actions and culturally defined uses for nature and natural resources. Epistemological questions, such as how people understand their relationship with their environment, are essential to a more sustainable development. They are not merely a desirable consideration in better planning, but, as we will see in chapter 7, are the very stuff of which environmental management could be made. To understand fully the importance of this culture–nature dialectic we need to begin with the way 'natural economy' was destroyed under the impact of colonial and post-colonial capitalist development.

The destruction of natural economy

The term 'natural economy' refers to small-scale societies prior to mercantilist contact in the sixteenth century which, as we saw in the last chapter, enlarged the compass of production and exchange, placing it on a global level. Natural economy thus refers to an historical society, rather than one that survives today. Such societies were pre-class societies, organized around kinship and using, on the whole, simple, integrated technologies, like those of the chinampas in Mexico. The units of economic production were relatively small and capital accumulation almost non-existent. Normally land was not privately owned, and access to publicly owned land was determined by considerations other than market forces, since, although market systems were often important, markets as we understand them were only partially developed.

Nevertheless natural economy does not imply a 'state of nature', since nature has never existed together with human societies without exhibiting the effects of human contact. Inequality certainly existed, for example, both between people of the same race and gender and those belonging to different group affiliations. However, the basis of stratification was not class, as in capitalist, industrial society, but status as determined by age, sex, religion and, to a limited degree, occupational specialization. These were societies characterized by 'mechanical solidarity' in Durkheim's sense (1964) in which monetization of exchanges was limited, commodity production was restricted to the limited needs of external trade and most, if not all, production was for use.

According to Rosa Luxemburg, whose treatise on the destruction of natural economy has influenced successive generations of thinkers (Luxemburg 1951), the complete destruction of natural economy was a gradual process which would occur through different stages. In the early stages the struggle was between capital and natural economy itself; later it would be waged between capital and commodity production. Ultimately the struggle would take a competitive form as different capitals attempted to take advantage of the remaining fruits of the accumulation process in the colonial state (Luxemburg 1951, 368). The penetration of natural economies was a direct result of the accumulation crisis in the developed countries, where demand for the products of capitalist industry could not rise further.

Events since 1945 would certainly lead one to question the continuing relevance of this assumption, but the force of Luxemburg's critique remains: the uses to which labour and natural resources are put, in the course of development, remove control from the local community. They also ensure that, as labour is 'freed' and market relations established, traditional relations with the environment are severed. The persistence of the struggle between what remains of natural economy and the capitalist market was attributed by Luxemburg to the greater opportunities for exploitation which contact with non-capitalist economies afforded: 'primitive conditions allow of a greater drive and of far more ruthless measures than could be tolerated under purely capitalist conditions . . . ' (Luxemburg 1951, 365). This view, that the transition to capitalism has been slowed down by the very mechanisms of exploitation that were established to achieve it, provided the basis for a lively debate between development theorists in the 1960s and 1970s (Roxborough 1979, de Janvry 1981, Goodman and Redclift 1981). Today we should add another factor, which has been given only scant attention in the past: the unbridled exploitation of labour and natural resources has produced contradictions of its own, among them the perceived need to manage the environment in harmony with the development process. Similarly, the unfettered working of commodity production, under conditions of resource scarcity or fragility, has contributed to what Bernstein has evocatively termed the 'simple reproduction squeeze' (Bernstein 1979) which can sometimes threaten the very conditions of accumulation.

The development of agriculture and the 'domestic community'

The discussion of natural economy should not lead us to ignore the important differences between societies which have yet to be brought fully within the compass of capitalist development. The simplest societies, at least from a technological viewpoint, such as those of gatherer-hunters, need to be distinguished from those of petty-commodity producers or peasants, for whom economic relations with the outside constitute a defining characteristic. The simplest societies are characterized by primitive plenty rather than scarcity, but they are not located in the Garden of Eden. It is a common misconception of tribal peoples that they are constantly nomadic, unattached to particular areas and aimlessly moving about the bush seeking new resources. In fact, there is a close attachment between tribal peoples and their lands. Their mobility is usually restricted to the area with which they are familiar and to which they have traditional rights. In Indonesia, for example, 'the only truly nomadic rural peoples are the displaced and landless peasantry from the central islands', most of whom have been effectively forced to transmigrate (Colchester 1986, 92).

In such societies very little time is spent on productive activities, since needs are defined by the existence of items in nature, and by the size and composition of the group. As little as four hours a day may be spent in hunting and gathering. The cycle for the reproduction of human energy is short. Subsistence foods from hunting and gathering do not keep well and must be consumed quickly. As a result there is little accumulation of the product, little surplus production. As Meillassoux puts it, 'the cycle of the transformation of food into energy is a daily one: virtually each day the producer exploits the energy he absorbed in the past few hours to produce what he needs to subsist for the hours to come' (Meillassoux 1981, 15). Unlike the situation in our own society, in simple-technology societies relations with the environment are intimate, continuous and involve a very short energy cycle.

The development of sedentary agriculture changes this. In peasant societies needs are no longer defined by the local group, market exchanges assume increasing importance, and the amount of work undertaken and surplus produced increase. The pattern of resource exploitation is closely linked to the exchanges which occur with the wider economic system. As commodity production increases, the needs of the simple commodity-producing unit (the household economy) are also defined in wider terms. In Meillassoux's 'self-sustaining' society of simple commodity producers, exchange relations with the outside world exist, but they still do not require a change in the relations of production. Increasingly it is what he calls the 'domestic community' which assumes importance, as longer-term commitments take precedence over the immediate requirements of food and shelter. Institutions like marriage and the dowry system come to dominate ceremonial life because 'they regulate not only the cohabitation of the married couple and their respective tasks, but the future positions of their anticipated offspring' (Meillassoux 1981, 39).

The development of sedentary agriculture and simple commodity production provides for a separation between 'productive' and 'non-productive' periods on the land. Consumption takes place continuously, but production is a much more discontinuous process, and one not as clearly related to meeting immediate needs as in hunting and gathering societies. Essential to the furtherance of sedentary agricultural production is the maintenance of stocks, of seeds and harvested products which will later serve to sustain the household during unproductive periods. These are the future surpluses which have proved attractive to capital in developing countries, but in many cases they existed independently of the development of metropolitan capitalism. It is not until capitalist relations of production are established that what Meillassoux describes as 'power in the mode of production' shifts from control over the means of human reproduction (subsistence goods and wives) to control over the means of production itself (Meillassoux 1981, 49). The distinction is an important one:

> The subsistence economy belongs . . . to capitalism's *sphere of circulation* to the extent that it provides it with labour-power and commodities but remains outside the capitalist *sphere of production* since capital is not invested in it, and the relations of production are domestic and not capitalist (Meillassoux 1981, 95).

How do these shifts from relatively autochthonous conditions to a much more integrated, specialized system of production and reproduction affect the relationship maintained between people and their environment? To answer this question we need to take account of several different facets of the process. Most economic anthropologists, following Marx's lead, have explored the relationship between nature and culture by concentrating on the notion of labour. As Eric Wolf has expressed it, in the course of development, 'humankind adapts to nature and transforms it for its own use through labour' (E. Wolf 1982, 74). Labour is therefore the key to the process, as far as understanding human societies is concerned. As different kinds of labour become interchangeable under capitalism, so we can determine the general trajectory of capitalism from the way in which labour is organized: commodities, their production and the social relations of production and reproduction become organized differently in different historical phases.

However, the natural environment also changes with these relations of production and reproduction. As we saw in chapter 5, the movement of commodities, to which Wolf refers at length in his excellent book, *Europe and the Peoples without History*, does not merely ensure that capitalism maintains its dynamic character, it also ensures that people enter into different relations with the environment. Eric Wolf argues that Marx himself recognized many of the complexities which still face us today when we consider the relationship between development and the environment:

> He was a materialist, believing in the primacy of material relationships as against the primacy of 'spirit' . . . For him production embraced at once the changing relations of humankind to nature, the social relations into which humans enter in the course of transforming nature, and the consequent transformations of human symbolic capability (E. Wolf 1982, 21).

It is interesting, I believe, that Wolf's reading of Marx, though an accurate one, is rarely commented on by anthropologists. Most, although not all, social-science writing about development has concentrated on the second of the observations attributed to Marx: the social relations which obtain in the course of transforming nature. To appreciate fully the importance of changes in the relations between human beings and nature, and what these imply for human symbolic capability, we need to address the epistemological question of how to begin to distance ourselves from our own preconceived view of the environment.

The nature of commodities and the commoditization of nature

One of the most interesting recent contributions to anthropological writing about capitalist development has been the work of Michael Taussig (1980). In his iconoclastic, provocative book *The Devil and Commodity Fetishism in South America*, Taussig reminds us that commodity production itself appears 'unnatural' to those in developing countries who make a first acquaintance with it:

> to their participants, all cultures tend to present [categories of thought] as if they were not social products but elemental, and immutable things. As soon as such categories are defined as natural, rather than as social, products, epistemology itself acts to conceal understanding of the social order. Our understanding, our experience, our explanations – all serve merely to ratify the conventions that sustain our sense of reality (1980, 4).

What Taussig is suggesting here applies particularly to knowledge about the environment and nature. As capitalism has developed and commodity-based exchange values have been substituted for use values, so we are confronted with an epistemological paradox. What was once 'nature' has become a product, while the process through which nature is transformed has become 'natural'. This inversion helps to explain the difficulty that agricultural scientists, for example, have experienced in making full use of traditional epistemologies, most of which use not only different categories of thought, but also different forms of explanation (Norgaard 1985a). As Taussig put it, '[in precapitalist societies] the meaning of capitalism will be subject to precapitalist meanings, and the conflict (with capitalism) expressed in such a confrontation will be one in which man is seen as the aim of production, and not production as the aim of man' (Taussig 1980, 11).

In his book Taussig sets out the contrast between precapitalist and capitalist societies' understanding of commodity production, in a series of what he terms 'positive and negative analogies' (129–39 *passim*). He is concerned principally with the 'baptism of money' performed in societies where exchange values are rapidly replacing use values. In these societies the destruction of nature implies a reduction in productive capacity. The impact of capitalism becomes increasingly important to the people who are charged with the task of increasing production while exceeding the limits posed by the ecological system:

> The problem facing the people in this culture is . . . how to explain and effect the inversion of these natural analogies, since the empirical fact of the matter is that production can be maintained and increased within the sphere of capitalist production (Taussig 1980, 134).

Taussig argues that the transformation of use values into exchange values, a transformation which implies an 'unnatural' response to nature, has convinced the participants in the process that capitalist relationships 'necessitate the agency of the devil'.

Taussig is principally concerned with the way in which capitalism has served to legitimate itself through appropriating 'naturalness'. The very term 'natural resources', as we saw in an earlier chapter, is part of this legitimation process. However, his argument has enormous importance for the way in which we conceive the development process, for the way that people relate to the environment is closely bound up with their understanding of social change. In the substitution of exchange values for use values, in the increasing separation of consumption from production time, in the definition of 'needs' which govern the livelihood strategies of the rural poor, it is futile to divorce peoples' practices from their beliefs. Whether they are willing or unwilling agents in the irrecoverable destruction of the environment, the rural poor are resocialized in the process, often lending credibility by their actions to the view that development is an inevitable, progressive process which is most successful when it is least sustainable. The conflict between behaviour consistent with market logic and that which can support sustainable livelihoods is seen clearly if we explore a specific case, that of colonization in the Bolivian Amazon.

Sustainability and accumulation on the Amazon frontier

In some respects the Amazonian frontier represents an exception to the general pattern of Latin American agriculture, in which even limited access to land is tightly controlled by governments and large landowning interests. Recent research has drawn attention to the ecological dangers of opening up a frontier so rapidly (Caufield 1984; Myers 1979). It has also increasingly concentrated on the social conflicts, largely between ranching interests, peasants and trans-national agro-industrial corporations which have reached consider-able intensity throughout the Amazon region (Foweraker 1981; Branford and Glock 1985; Plumwood and Routley 1982; Barbira-Scazzochio 1980). Rather less attention has been given to the experiences of practical conservation which forms part of the indigenous knowledge systems which environmental planners could make use of (McNeely and Pitt 1985). Similarly, little attention has been focused on the opportunity afforded indigenous people to 'develop' alternative cultural and ecological systems through closer political links (Dunbar Ortiz 1984).

As the frontier is opened up, access to land is relatively easy and, initially, is determined by factors other than the ability to pay for it. The availability of land and the shortage of labour enable simple commodity producers to establish themselves. The quality of the natural resource base is highest in tropical frontier areas at the geographical 'margin', where land is brought into cultivation for the first time. The potential exists for self-provisioning (made more necessary by the distance from urban markets) to be combined with small-scale crop production in both staple food crops and export crops. The history of frontier situations in Latin America, however, suggests that this potential is seldom realized. Land soon becomes monopolized by large farmers and agro-industrial interests able to exploit the situation which colonization facilitates: a cheap land market, a family labour force free from institutional restraint and the inducement represented by an ever expanding land frontier. The profitability of agriculture ensures that accumulation occurs within the frontier zone, the level of land concentration increasing the further one is from the conversion of virgin forest.

During the last decade attempts have been made to devise appropriate agricultural production systems for humid tropical environments which recognize the ecological fragility of these areas as well as the need for compatibility between ecological and social systems. Norgaard, in an interesting paper, demonstrates how the development of the Brazilian Amazon has imposed a technology and infrastructure which correspond neither to ecological conditions nor to the experiences and skills of the colonizing population (Norgaard 1984a). In similar vein Moran sets out the conditions under which colonization in Amazonia could help build a sustainable agriculture, notably by distinguishing carefully between localized soil types and seeking to make more use of indigenous knowledge (Moran 1984).

One of the principal obstacles to sustainable agriculture being practised more widely in Latin America's humid tropics is the process of land accumulation, through which land and natural resources on the frontier become fully integrated within developing capitalist economies. The impetus behind this process usually originates at the international level, where the demand for primary products has stimulated the search for new zones of production. In Brazil, for example, the 'spread of the coffee frontier into São Paulo and then Parana was determined by the level of world demand for coffee, which in turn affected the profitability of constructing transport links to extend the pioneer fringe' (Katzman 1975, 283). In some cases frontier expansion was undertaken by a land-holding élite, for whom land represented an attractive investment opportunity (São Paulo, Brazil); in other cases frontiers were deliberately and successfully developed for small farmers (San Ramon in Costa Rica; Parana in Brazil). The social agencies through which frontier expansion has been achieved have varied, but the inexorable pressure to open up more land to the market has remained a feature of Latin American development.

Recent theoretical discussion of Latin America's tropical frontier has focused on the roles of primitive accumulation and large-scale agro-industrial capital. Among the most influential writers, Oliveira (1972) has argued that the frontier serves as a means of mobilizing the agricultural surplus through the mechanism of primitive accumulation and the exclusion of labour from permanent access to the land. The 'elaboration of peripheries' has been impelled by spontaneous colonization and has, in the early stages, enabled the conditions of existence for a peasant economy to be established. Gradually land has become accumulated in fewer hands and the *latifundia* system of production has been reproduced on successive agricultural frontiers. Other writers, notably Velho (1976), have pointed to the subordination of the peasant mode of production within the capitalist process of permanent-surplus appropriation. In Foweraker's (1981) view the articulation of different modes plays an important role in the provision of a marketed surplus of staple foodstuffs to urban centres. In his opinion the frontier cycle, denominating a transition from pioneer peasant agriculture to capitalist production, although geographically marginal, is central to the conditions of accumulation in Brazilian agriculture.

This argument is contested by a number of authors (Goodman 1984; Sorj and Pompermayer 1983; Sawyer 1979) who have emphasized the increasingly important role of large-scale capital in frontier expansion. Corporate capital 'has used heavily subsidised investment credits to purchase huge tracts of public land at nominal prices for cattle-ranching and private land settlement' (Goodman 1984, 51). In the view of these critics the reproduction of peasant economy under frontier conditions is problematical. Frontier expansion does not *require* a subordinate peasantry. Indeed conflicts between small and large landowners arising from frontier expansion may even jeopardize the political control exercised by state agencies and agribusiness corporations. Conflicts between social classes, by entering the political sphere, may call into question the pattern of development produced by state incentives.

Exploring another aspect of frontier expansion and concentrating specifically on the livestock economy, Hecht (1985) has argued that environmental degradation in eastern Amazonia is related not to the productivity of land itself, but to the role of land in inflationary economies. Thus ranching is a 'means of acquiring large areas (and the institutional rents associated with them), the stimulating effect of the physical opening of the agricultural frontiers in certain industrial sectors of the economy, and the role of large government subsidies in the creation of land markets and speculation' (Hecht 1984, 38). Some land uses and technologies, such as ranching, serve important political and ideological functions, by helping to ensure the political conditions for accumulation. Since the productivity of the land is less important than its value as a commodity, land degradation is the logical outcome of frontier development. This line of argument clearly identifies a process of accumulation which, while linked to urban and industrial capital, is essentially a means to the appreciation of capital values rather than the supply of a wage good, food.

Most of the debate in the literature relates to Brazilian frontier experience, but the process of accumulation behind frontier expansion can also be observed in the Bolivian province of Santa Cruz (Jorgensen 1973; Gill 1985). Once again, the incorporation of new land has contributed to environmental problems by closing certain agricultural options, and ultimately dispossessing many of those who initially colonize the frontier. However, in the Bolivian case the economic viability of capital-intensive agriculture has recently been called into question, largely as a consequence of the country's bankruptcy, and the search for alternative, more sustainable systems of cultivation has gained ground.

The Bolivian experience of colonizing the Amazon frontier has been particularly interesting, but much of the discussion has been restricted to unpublished papers written by development workers with non-governmental organizations and agricultural economists, like those associated with the British Tropical Agricultural Mission (BTAM) in Santa Cruz (Maxwell 1979; Lawrence-Jones 1984). After 1952 land reform in highland Bolivia helped to create increasing demographic pressure on the limited resource base. Many *campesinos* migrated to the Eastern Lowlands, initially to work on the sugar and cotton harvests, but often with the intention of staking a claim to the expanding Amazon frontier. State policies encouraged this process with the objective of boosting agricultural production, attracting workers to a labour-scarce region and providing land to potentially militant peasants (Gill 1985). The buoyancy of the Santa Cruz economy rested increasingly on the government income derived from oil production. During the 1960s there was also a limited official colonization of the region. Between 1962 and 1971 it was planned to settle almost half a million people in three colonization projects sponsored by the Inter American Development Bank (Schuurman 1979). In fact only a small proportion received official assistance. The large majority found their own way to the frontier, especially the northern colonization zones, after working in the *zafra* (sugar harvest) near Santa Cruz.

Land concentration in the area near Santa Cruz was a feature of the 1960s and 1970s, much of it spent under military government. Large public land grants were made and these, together with the encroachment on to the land of peasant settlers, ensured that agro-industrial expansion proceeded rapidly. Today, as table 6.1 shows, 3 per cent of the largest holdings account for over half the titled land. Deprived of land in the integrated zone of highly commercial agriculture, the small settler population had to push further and further into the jungle. At the same time the large commercial farmers were introducing technological changes which had a profound effect on the labour process, substituting proletarian for family labour in successive 'frontier' stages. The increasing power of the commercial farmers, together with entrepreneurs, government personnel and speculators who were attracted to the zone, was represented through powerful producer interest groups

Table 6.1 The occupation of land in Santa Cruz, Bolivia, 1981

Farm size	No. of titles	(%)	Extension (ha)	(%)
0–500 ha	43,017	97.0	4,196,462	45.5
500–2,000 ha	632	1.4	691,074	7.5
2,000–10,000 ha	659	1.5	2,667,534	28.9
10,000 ha +	70	0.1	1,676,735	18.1
	44,378	100.0	9,231,805	100.0

Notes:
1 These figures have been calculated from two main sources of data: land registration documents and land distribution under the 1953 Agrarian Reform Law. The data refers to titled land only, i.e. it excludes most of the land in the colonization frontier. It thus underrepresents private occupation of land under 50 ha in size.
2 The figures refer to land titles rather than land owners. They fail to reflect multiple ownership of land titles by the same person or family. The concentration of land ownership for which titles exist is likely to be greater than the figures suggest.
Source: The principal source of data on which these calculations are based is: *Tierra, estructura productiva y poder en Santa Cruz* (1983) La Paz, Grupo de Estudios Andres Ibanez.

(BTAM 1985, annex 17). Many became embroiled in highly lucrative cocaine traffic and contraband activities which provided an additional speculative activity in the province.

The effect of these processes of land concentration and capital accumulation was not to induce generalized proletarianization, since the existence of land on the 'cutting frontier' enabled many *colonistas* to combine the small-scale production of rice (for the market and subsistence) with wage employment. As Gill has demonstrated, the livelihood strategies adopted by the settler population were an adaptation to the existence of both capital-intensive agriculture and relatively easy access to 'new' land on the frontier. Thus, 'while the better off settlers struggled to defend their position as small-scale producers above all else, the response of the semi-proletariat reflected the diverse strategies of wage labourer and subsistence cultivation which they adopted at different times in order to survive' (Gill 1985, 17). By 1984 the rate of inflation in Bolivia had reached 2700 per cent, and the real value of wages had become so eroded that many semi-proletarians returned their attention to small-scale subsistence cultivation or invaded new land on the frontier. The attachment of colonists to land was increased by the decline of the commercial sector, which survived through heavily subsidized inputs made easier by parallel exchange rates.

Farming systems in Santa Cruz's 'colonies'

Farming-systems research in areas like Santa Cruz needs to find solutions to the problems presented by the fallow (*barbecho*) system, in the transition from slash-and-burn agriculture to permanent cultivation. What Maxwell (1979) has termed the '*barbecho* crisis' consists of a number of closely interrelated problems: the reduction of soil fertility over a period of four or five years; the rapid regrowth of weeds on fallow land; and the acute shortage of labour to tackle these problems under frontier conditions.

The avenues of escape from these problems have usually taken one of two forms. The solution favoured by most outsiders until recently was for colonizer farmers to 'destump' the land and introduce tractors for one of a number of farm-management activities (usually sowing and harvesting). Rice would then be grown on the land under cultivation. This would provide a regular source of income, and the land would appreciate in value, enabling small colonizers to sell it to larger operators for their own benefit.

Such practices have negative effects, however. The task of destumping and levelling with tractors is costly and frequently results in the farmer's indebtedness. The ecological effects are also damaging, as soils become eroded with the removal of vegetation cover. There are also physical difficulties in transporting heavy machinery between plots which are frequently 20 or 30 kilometres apart, and in appalling weather conditions during the heavy rains. Specialization in rice production leaves the small farmer further exposed. The rice price is currently supported, since large mechanized rice producers form an influential interest group at the regional level. There is, however, a real prospect that these subsidies will not continue, especially as Bolivia's oil revenues are gradually exhausted. The non-payment of Bolivia's foreign debt and the failure to achieve an agreement with the IMF for a standby arrangement have resulted in the virtual cessation of international-agency project finance to Bolivia. Finally, access to agricultural credit in the area is made more difficult by the confusion surrounding land titles. The modernization option is thus neither environmentally sound nor (for most farmers) politically realistic.

The second path of escape from the problems presented by 'the *barbecho* crisis', is the sale or abandonment of land and the colonizer's return to the *monte alta* (virgin forest). This is a course of action undertaken by many households, but it is hardly a solution to the poor farmer's problems. The removal of tropical forest is time-consuming and physically exhausting. Few men over the age of 40 even attempt it, and fewer still return to try a third time. Land is relatively abundant on the cutting frontier in Santa Cruz, but for that reason alone it cannot be sold at a price which gives an adequate return to the colonizer. Necessity, rather than choice, propels people back to the frontier.

There is also a third solution to the '*barbecho* crisis'. This is to provide a more permanent, sustainable farming system through combining annual crops (maize and rice), perennials (cocoa, coffee and plantains), small livestock production (chickens, sheep and pigs) with legumes which enrich the soil and prevent weed regrowth. The objective of this system is to address the needs of small farmers, without taking an unacceptable toll of the environment. First, the reduction of weed infestation and the adoption of perennials can ease labour shortages. Second, intercropping with several staples, as well as perennials, can help maintain yields without imperilling essential cash income from marketed crops. Third, diversifying production helps to spread risks and enables small farmers to experiment with new crops. Fourth, lowering the dependence on bought inputs, especially oil-based fertilizers, can help the farmer assume greater management responsibilities. Small livestock production, which thrives informally in the colonization zones, is a key element here. Pigs and chickens are fed with household waste and, some-what surprisingly, sheep are kept very successfully on the frontier by women colonizers who came originally from the *sierra* regions. The ultimate aim of this farming system is to increase equity and sustainability without reducing the productivity of the system.

However, the successful introduction of farming systems of this kind on the Bolivian frontier requires more than appropriate agricultural technologies, important as these are. The obstacles are often structural, in the form of incentives to adopt unsustainable, speculative development strategies. Eastern Bolivia is one of the centres of the traffic in *coca*, the leaf from which cocaine is derived. One night's work treading coca leaves pays a day labourer the equivalent of a week's wages in agriculture. Santa Cruz is also the centre of the contraband traffic across the Brazilian border. For many people the prospect of easy money, whether in coca, contraband or land speculation, is much more attractive than embarking on new systems of production within a longer time-horizon. In addition, not everybody on the Bolivian frontier is involved exclusively in farming. Many individuals and most households generate income from other activities: petty trade, wage labour, trucking and transportation. What is sustainable agronomically and ecologically may thus prove unsustainable for farm households; it may prove to be *socially* unsustainable.

Household survival strategies and sustainable development

If technical solutions exist to the problem of establishing sustainable farming systems on the Amazon frontier, what factors will determine whether colonista farmers are willing and able to put these systems to use? In answering this question we are widening the compass of farming-systems research, by considering the demand component in the development of agricultural technology as well as the supply component. This is not simply a matter of taking farming systems research further 'upstream'; it means considering farming systems themselves as one of a number of strategies open to households on the frontier. We need to know under what conditions colonistas will seek to invest their time and energies in the systems which are currently being developed.

The demographic cycle and rural settlements

One of the principal factors influencing colonizers in their selection of a farming system is the stage that has been reached in the development of their domestic cycle. Older men usually prefer to work fallow land than virgin forest because, although the soils are less fertile, it is less physical effort clearing the land. In the settlement of San Julian, which is today some distance from the cutting frontier, a survey carried out in 1983 found that two-thirds of the adults were married, and almost half of these people had between four and six dependants (Taylor 1983). Households such as these are already settled in most respects, including the possession of land titles and permanent housing. Families usually possess 50 hectares of land organized on a 'piano key' arrangement, off one of the unsurfaced lateral roads which run at right angles to the roads that penetrate the area. The settlement or *nucleo* typically consists of 2000 hectares of land, divided between about 40 families, and extending outwards from the basic urban services: a water well, small school, football field and store. Not all nucleos in the zone are as well planned, but those that are provide a likely target population for agricultural development.

The sexual division of labour within the household

To appreciate fully the way in which the population of the 'colonies' adapts to a new environment, we need to consider the sexual division of labour within the household. While it is true that men do most of the tree-felling, the accompanying tasks are often undertaken by women (burning scrub, weeding and planting). The care of animals is largely in the hands of the women, especially the sheep which form an essential part of the family's livelihood. The sheep from several families are grazed together and their wool, once woven into cloth, provides an important source of cash income. The evidence suggests that sheep are entirely complementary to the rest of the household's activities, and their care is not labour-intensive. Other animals, such as chickens and pigs, are fed with household waste and thus are usually kept close to the house, rather than on the land holdings, which might be some distance away.

Where cash crops such as coffee have been introduced, it has been the women who have carried out much of the work of tending the plants, weeding, etc. Nevertheless, women are rarely given economic control of this activity, since the agricultural credit and marketing is in the hands of the men. A farming system involving perennial-crop production or small-scale animal production should concentrate on the rewards and investments of women (especially in time) rather than simply assuming that men, as household heads, take decisions with the welfare of all family members in mind. From a practical standpoint, it is unlikely that an effective balance will be found between staple and cash crops unless women participate more fully in the control of resources, as the experience of PLADERVE, a project based in San Ignacio, has already indicated (BTAM 1984, annex 14).

Community-based workgroups

It has been assumed by most large development agencies that colonizers on the frontier act individualistically. The evidence, such as it is, does not readily support this assumption. Conversations with the staff of a number of the non-governmental organizations operating in the zone suggest that many colonizers have migrated from highland communities to specific settlement areas, that most have some experience of working together in the sugar harvest and that a majority have strong ethnic-group affiliations. For example, reciprocal labour arrangements between individuals (*ayni* in Quechua) exist. Similarly, communal labour arrangements are also important in the zone, involving labour exchanges (food, drink and the provision of coca to those who participate). This is the *mink'a* known to anthropologists in the Sierra.

Evidence also exists of larger-scale activity. In 1981 a major road repair project was organized using mink'a labour, drawing in total on fourteen nucleos in the San Julian area. Over three hundred people worked for two weeks to raise a stretch of road 100 metres long above the floodline. Such activity is not uncommon, and the local non-governmental organizations closest to the colonists frequently report activities which suggest that the frontier is not permeated with unchecked individualism. The appearance of 'disorganization' (like the findings documented for urban settlers throughout Latin America) frequently tells us more about the perceptions of outside observers than about the social structure of newly colonized areas.

Work groups of the kind found in San Julian illustrate the considerable sacrifices which colonists make in order to maintain a poor and deteriorating infrastructure, without which they cannot get their perishable produce to market. In the Chane-Piray zone there are also reports of enforced participation by colonists in road-clearing operations. Non-compliance with these obligations resulted in the threat of expulsion for colonist families. Mink'a, which began as an indigenous response to the new conditions of the colonies, has become an instrument of state coercion available to the dominant social groups.

These examples should serve to remind us that the colonizers are heterogeneous in their social and cultural composition, degrees of internal differentiation, the level of their insertion into the regional economy and their articulation with different fractions of agro-industrial capital, as well as the particular ecological constraints (such as vulnerability to flooding) which need to be overcome. Community-based workgroups are not a solution to every problem encountered by colonists, but they are often an effective and necessary institution, part of the survival strategies from which colonists draw considerable strength in the face of a difficult physical environment and a vulnerable structural position.

The prospects for sustainable farming systems on the Bolivian frontier

There is frequently a speculative, adventitious quality to the way land frontiers are opened up. The Bolivian Amazon is no exception. Relatively easy access to new land and the existence of a market for land that has been cleared encourage colonizers to place short-term interests above long-term sustainability. At the same time few farmers will make a commitment to perennial crops with a long growing cycle, such as cocoa or coffee, when financial instability is endemic. Finally, alternative land and labour uses are extremely attractive: petty trading, cocoa, contraband-related activities. Unless structural policies are introduced which provide incentives for long-term sustainable resource uses, they are unlikely to be adopted on the scale required.

Many of the generalizations about the development of the Amazon frontier are borne out by the experience of Santa Cruz. There is considerable evidence of land concentration and of political leverage on the part of a new agro-industrial class. The rapid appreciation of land values plays an important part in this accumulation process, but so too does the subsidized production of commercial crops, including one (rice) that is an important food staple. The Eastern Lowland provinces have played a vital strategic role within the development of Bolivia in the last two or three decades. In the case of the northern colonies of Santa Cruz, this role has produced a particularly striking contrast between the rapid penetration of market forces and the process of environmental degradation.

Today the economic viability of the Bolivian tropical frontier is in doubt. Apart from the enormity of the environmental problems, which have been documented by the Regional Development Organisation (CORDECRUZ), economic recession has begun to affect Santa Cruz, as oil earnings fall and the national economy is locked in bankruptcy. As is the case in other Amazonian countries, the Bolivian state is heavily involved in underwriting the agricultural economy of Santa Cruz but, as yet, is apparently unwilling to implement long-term policies favourable to sustainable farming systems.

The obstacles to establishing sustainable rural development do not make it any less important an objective, however. Evidence exists from various parts of Latin America that agricultural development can take place without environmental degradation (Sanders and Lynam 1981; Hecht, Anderson and May 1985; Goodland 1981; Stanley 1984; Norgaard 1981). There is no shortage of information about alternative farming systems which, used selectively (and poor farmers usually act selectively), could help improve the livelihoods of the environmentally and socially marginal. At the same time the process of land accumulation and capital penetration in areas like the Bolivian Amazon reduce the likelihood that technical achievements will be institutionalized. The agricultural frontier can often be negotiated by speculative risk-taking, rather than the kind of risk-avoidance which has been a central element in peasant livelihood strategies elsewhere. In the absence of policies to encourage colonizers to adopt more sustainable farming systems, and which recognize that livelihood strategies spanning several activities frequently imply short-term time horizons in each of them, it is unlikely that long-term, environmental damage will be reversed. The strategies which the colonizer population has developed to minimize the harm which the market has caused will lead to natural-resource degradation and reduced life-chances for the next generation.

The immediate task, then, is to identify the components of people's livelihoods that exist outside the boundaries of the farm, but which carry implications for agronomic research. For many households farming is not an exclusive activity, but mobility between activities is essential to the survival of the household. 'Systems' also include labour, notably that of women, which is not accorded market value, and resource uses (the provision of energy, water, shelter and food) which lie within the domestic sphere and are rarely considered as part of the environmental picture.

This chapter has examined the effects of changes in the environment at the local level. It has argued that we need to be more aware of the relationship which links peoples' understanding of the environment with their behaviour. Finally, it was suggested that the livelihood strategies which people pursue, in ecologically fragile areas like the Amazon, provide important pointers to the kind of technologies that are most appropriate on social as well as ecological grounds. In the next chapter we explore more fully the implications for social action which are implied by alternative epistemological positions.

7

Environmental management and social movements

Previous chapters have discussed what is happening to the environment in developing countries in the light of historical experience, and the kinds of structural processes within the international economy which have served to alter the relationship between developed and less developed regions.

In discussing the transformation of the environment it was necessary to examine sustainable and unsustainable approaches to development within the context of peoples' choices and interests, as well as ecological principles. In this chapter the argument turns full circle and we examine the principles and methods of environmental management, beginning with the human agent as the manager of the environment. When we consider the principles and methods of environmental management as practised in the developed countries, it is evident that most types of intervention are neither particularly effective nor grounded in an objective scientific approach, as is claimed by some practitioners. Environmental management is usually a responsive set of techniques rather than a framework for implementing policy. It also becomes clear that what we mean by environmental management often ignores or devalues the experience of poor people in the developing countries, those who are usually closest to the problems. We are often in danger of importing solutions to environmental problems from the experience of developed countries, using methodologies and an epistemology which is of little relevance to different circumstances. To begin to achieve success at implementing workable environmental policies in the South we must first unlearn much of what we know about conservation and the environment in the developed countries.

Consider the three main objectives stated in the World Conservation Strategy (1980): the maintenance of essential ecological processes;

the preservation of genetic diversity, and the sustainable utilization of natural resources.

To achieve these objectives clearly requires social and economic interventions of various kinds. Even in the privileged context of a national park or biosphere reserve, these interventions are likely to succeed only if the environment is accorded as much priority as other variables in the development process, such as the growth of marketable commodities, or the need to meet higher levels of personal consumption. The implications of quite small changes in the way the environment is managed are often enormous, and the achievement of modest environmental objectives is difficult. Eco-development, as presently practised, falls short of being the prescriptive, workable framework for planning that its advocates would like it to be (Bartelmus 1986, Glaeser 1979).

The type of intervention open to environmental planners in developed countries varies according to the type of conflict over the environment. There are broadly three types of conflict: in the sphere of production, in the sphere of consumption and in the sphere of nature. In the sphere of production, 'intervention has been necessary to overcome market failure in the regulation of externality effects' (Blowers 1985, 11). Within industrial society capitalist firms seek to avoid the externalities of other producers, and to secure their access to essential natural resources. Since the late nineteenth century the establishment of government resource agencies has sought to mediate conflicts within this area. At the spatial level one of the most important approaches adopted has been land-use planning which attempts to zone areas in which resources could be exploited. Methodologies have been developed to try to ensure that only land which is capable of being developed in certain ways is permitted for development. As we shall see, this kind of land-use planning exercise has been influential in the designation of biosphere reserves in both developed and developing countries.

Another type of intervention is in the sphere of nature conservation, where conflicts have been generated over the effects of modern agriculture on flora, fauna and their habitat. These conflicts, as we will see, often concern the transformation of natural species into commodities, with the usual rights of ownership and control corresponding to commodities. A fundamentalist view of the rights that we should assign to nature is the hallmark of the Deep Ecology position discussed in chapter 3. Most discussion in the developed countries turns on the conservation of critical habitats in which rare or endangered species can exist but, as we shall see, the issues are frequently far wider in developing countries, where the threat to the species carries implications for the global gene pool itself.

The conflicts that arise in the sphere of consumption alone are familiar to most of us in the developed countries, but of much less importance in most parts of the South. In developed countries the objective of environmental groups is frequently to secure, or preserve, access to pleasant and safe residential and recreational areas. Providing environments in which people can enjoy their leisure is one of the objectives of amenity planning, which may need to mediate conflict between different social groups whose behaviour would otherwise be impossible to reconcile, such as ramblers and farmers, or hunters and conservationists. In some cases production-oriented groups, like farmers or foresters, are constrained from exercising certain powers because of an acknowledged public interest in amenity.

These different forms of intervention and planning are hardly radical. In the case of land-use controls, for example, 'development may be permitted or refused . . . but refusal does not prevent development since it may be allowed on appeal, as the result of resubmission or may simply be diverted elsewhere' (Blowers 1985, 11). The effect of land-use controls, then, is to influence the location of development, not its occurrence. Other kinds of environmental intervention are equally modest in their effects. The control of pollution, for example, is usually only possible with the tacit approval of the polluters (Sandbach 1980). A balance is struck between economic and environmental considerations which is frequently favourable to economic forces, especially when the economy is depressed and the call for jobs and economic growth is difficult for politicians to resist. In the United Kingdom, for example, most environmental controls involve administrative discretion by the government inspectorate, and the enforcement of environmental controls frequently leaves pollution havens in areas less subject to effective control.

These methods of intervention were designed to facilitate, rather than seriously curtail, the production activities which are central to industrial economies. Most environmental responses to the existence of conflict have been what Holling (1978) calls 'protective and reactive responses'. Each problem, and its separate occurrence, is treated as if it were unique, as if the environmental consequences could be separated from the social and economic ones. The method employed is *ex post* rather than *ex ante*. As Holling (1978, 6) says:

> The result of simple reactive response is therefore intolerable. How can we know what to measure for baseline information or assessment if the detailed character of the policy or development is not revealed until it has largely crystalised? The tendency is to measure everything, hence producing the indigestible tomes typical of many environmental impact statements. More time and effort are spent in measuring what is, rather than in projecting what is likely to be, or could be made to be. Static and confused description replaces anticipation and clear prescription of alternatives.

It is extremely important, then, that we do not exaggerate the importance of environmental planning and management in our own societies. The methods employed – land-use planning, the costing of environmental losses, the development of priorities for species conservation – are reactive responses which have been developed to deal with the uncomfortable consequences of economic growth. More effort usually goes into making inventories and gathering statistics than into the redirection of the development process. Important areas like the appraisal of technologies and their environmental effects are rarely invoked to curtail development. Most actual development implies short-term efficiency for a small number of people. Although some writers regard environmental management as a panacea, even for the industrialized societies (Rosenbaum 1973), in reality progress has been slow. Even in the United States, where much of the technical framework for soil and water conservation originated, problems like soil erosion have been little influenced by policy. Cook (1983) makes the point that despite land-use planning in the United States, land that is unsuitable for arable cultivation has come under the plough. The ability to undertake technically sophisticated land-use assessment does not mean that effective policy measures are taken to ensure that land is used in the most appropriate way. The production of surpluses on unsuitable land – an estimated 27 million acres of land in the United States – is a response to market signals and has little to do with land-use planning. Environmental management, imbued with the contradictions that afflict all management sciences, represents an attempt to mediate the contradictions of industrialized society by minimizing the social costs of conflict. The need to 'manage' arises when 'traditional understandings and modes of social action and cultural formation have begun to prove inadequate, and the problem of controlling the *subjective* coordination and development of society presents itself' (Hales 1986, 84). In discussing environmental management we need therefore to be aware of the ideological significance of managerialist approaches as well as their limited success and relatively narrow application.

A managerialist view of the environment corresponds with what O'Riordan (1981) has termed a 'technocentric' (rather than an 'ecocentric') perspective. The assumption is that an optimum balance of natural-resource uses can be found, which can combine productivity with conservation goals in agriculture and forestry. Those who are convinced of the technical feasibility of identifying optimum resource uses also assume that long-term interests in the environment are convergent. Moral persuasion may be required to persuade groups of their shared long-term interest, but the ability to mount the necessary persuasion is governed by the level of funding available – there is no objective reason why the long-term interest of the biosphere should not be recognized by everyone.

This kind of argument falls on a number of counts. First, it is impossible to arrive at the optimum mix of resource uses without preconceived, value-based criteria. Once you agree about the kind of society you want, then agreement about environmental goals may not be difficult, but a social consensus does not exist. Second, as we have seen, most environmental management is corrective, rather than directive. Corrective interventions are most successful when they prevent something from happening, but this also makes them difficult to evaluate. As long as environmental management is conceived principally in terms of avoiding the negative externalities of the development process, it is inevitably difficult to assess. In addition, since much environmental planning is concerned with relocating development, rather than preventing it, the real damage to the environment often occurs in less well-protected areas far from the conservation zone. This is a very real problem in developing countries. Finally, environmental problems have characteristics which make for relatively facile diagnosis but difficult solutions. It is much easier to establish what has happened than why it has happened, because of the complexity of variables at work in establishing causation, the relevant time horizons and the correct environmental parameters. For all these reasons environmental management is not only unable to initiate radical action, it is also unable to make much impact on the problems that exist. The 'facade of technical objectivity' obscures the fact that environmental management is in its infancy. It is still concerned with 'techniques rather than policies' (Blaikie 1985). Despite appearances to the contrary, environmental planning is therefore often toothless. Examples are not difficult to find. Under the action programme of the European Economic Community for the environment most environmental planning 'has no binding effect on Member States. (They) may well adopt it somewhat cynically for this reason alone. Good environmental rhetoric has no political costs' (Sandbrook 1982, 331).

Biosphere reserves: environmental management in action

Many of the principles at work in the environmental management of developed countries have become internationalized and are currently being applied in the South. A good example is that of the 'biosphere reserve', a concept which is wider than that of a national park, since biosphere reserves are intended to meet global conservation objectives. Within the industrialized world the number of officially designated reserves of this kind has increased in recent years. Table 7.1 sets out the number of such sites, and their total extension, within the OECD countries. In countries like the United States and Australia much of the land included in biosphere reserves is wilderness. According to some commentators 'the wilderness values discovered among national park users support the idea of Nature as a safety valve' (Jeans 1983). One of the principal objectives of reserves is therefore to enable the public to have access to space which cannot be developed.

The situation in most developing countries is significantly different. The idea of protected areas persists, but the concept of biosphere reserves leaves amenity issues very much to one side. According to Batisse (1985) biosphere reserves 'combine Nature Conservation with scientific research, environmental monitoring, training, demonstration and environmental education' (p. 17). Today there are 243 biosphere reserves in a total of 65 countries,

Table 7.1 Biosphere reserves, selected industrialized countries

	Number of sites	*Total size (km²)*
Canada	2	581
USA	41	108,111
Japan	4	1,160
Australia	12	47,231
Austria	4	276
Denmark[1]	1	700,000
France	4	3,448
Germany	1	131
Greece	2	88
Ireland	2	88
Italy	3	38
Norway[2]	1	15,550
Portugal	1	4
Spain	9	4,250
Switzerland	1	169
UK	13	443
Yugoslavia	2	3,500

Notes:
This table contains data about the numbers of biosphere reserves and their total size for selected countries. The concept of the biosphere reserve combines nature conservation with scientific research, environmental monitoring, environmental training and education and local participation.
[1] Data refer to the Northeast Greenland National Park
[2] Data refer to the Northeast Svalbard Nature Reserve.
Source: OECD 1985

both developed and developing. The basic idea is that representative areas of ecological importance should be linked together through a network of information. Each reserve would contain at its core a minimally disturbed ecosystem. The preservation of genetic richness in this system would be able to withstand the processes of attrition which have affected protected areas in most parts of the world.

The compromise contained in the concept of biosphere reserve is that 'a balance must be achieved between managing the important environmental values of the world's more important remaining wildlands on the one hand, and converting some of them to more intensive uses on the other' (Goodland 1985, 4). The intention is that the economic development process should be able to incorporate a concern with wildlife and habitats. To this end reserves consist of several zones around a central protected core. Outside the core area

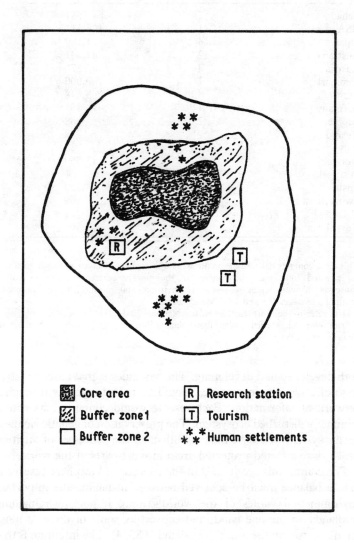

Figure 7.1 A typical biosphere reserve showing the zones

buffer zones exist in which only limited agricultural activities are allowed and research stations are located. Minimum disturbance of the buffer zones enables the most protected areas to survive intact, but an accommodation has to be found with the local human populations who habitually use the reserves. A typical biosphere reserve is illustrated in figure 7.1.

The notion of a core area in tropical zones for conservation corresponds with what is known as Pleistocene Refuge Theory, a suggestive approach to prioritizing zones for conservation. According to the theory, during the Pleistocene period the tropical forests receded into areas with higher precipitation, today's tropical rain forests. It is logical to suppose that these areas are best able to provide the genetic material for a recolonization of areas disturbed by human population. Although the reasons for the survival of tropical forest areas from prehistory are disputed, the value of preserving these areas is not. The scale of deforestation, especially in the tropics, limits the amount of research time in which the flora and fauna of these regions can be categorized and evaluated. In addition, the genetic resources of tropical forest areas are of considerable commercial importance, as well as constituting the world's principal gene pool. As we shall see in the next chapter, the inadequacy of gene banks as a way of storing genetic material lends additional weight to the biosphere reserve as a banking system. Seed storage does not work for many plants, especially propagated plants like potatoes, cassava and members of the orchid family. Many tropical plants can be conserved outside the wild only with great difficulty. Gene banks, unlike reserve areas, 'freeze evolution', since they do not allow species to adapt to new ecological circumstances, predators and diseases.

The ecological principles behind biosphere reserves, however, are only one side of the equation. The actual protection of areas, once they have been established on paper, has not fared so well. The World Bank is aware of the need to ensure 'effective management "on the ground" not simply on paper' (Goodland 1985, 11). The Bank's Office of Environmental Affairs notes, however, that 'the mere declaration of intent to protect wildlands or wildlife . . . does not ensure effective management unless specific supporting measures are implemented' (Goodland 1985, 12). In developing countries the pressures on biosphere reserves from 'the poverty of the poor and the greed of the rich' (Goodland 1985, 3) is much more formidable than the pressures experienced in developed countries, where most of the reserves' users are likely to be the strongest supporters of the conservation idea. The differences between national parks in developed and developing countries are summarized to take account of these factors in table 7.2.

Table 7.2 National parks in developed and developing countries

Developed countries	Developing countries
Land is owned – privately or by government	Most land is part of 'the Commons'
Land is of poor agricultural value	Poor land but rich land market
Amenity use – public access to nature	Little public access
Limited farming population – under protected status	Indigenous population – under threat

The essential differences between reserve areas in developed and developing countries is illustrated by Fearnside (1985) in an article about the Brazilian government's role in the development of the Amazon. Fearnside shows that there are two broad classes of activity shaping the Amazon landscape: the first is extractive and includes mines, dams, wood technology and logging, agricultural and ranching programmes; the second type of activity includes research on potentially sustainable forest-management systems, basic ecological research and the demarcation of Amerindian reserves. He concludes that 'the preference for projects in the first category is obvious from the resources allotted to developments in the first group . . . by Brazil's Planning Secretariat (SERPLAN)' (Fearnside 1985, 243). The existence of new scientific inputs cannot correct this situation, since the decision to give priority to non-renewable development is a political decision. Furthermore, the input of scientific research, except in a few high-priority conservation zones, does not enter into the decision-making process at a sufficiently early stage to affect the basic structure or existence of the projects in question. The relatively low priority given to parks and reserves by the Brazilian government is also evident from the minimal budget and staff allocated to the Brazilian Institute for Forestry Development (IBDF). Biosphere reserves provide a good example of the limitations of a model derived from the experience of countries in which environmental management is very far from being a livelihood struggle. As we have seen, the transformation of the environment in developing countries has given rise to social struggles that are also livelihood struggles in which a concern with the distribution of resources is secondary to creating the conditions of existence for further accumulation.

Environmental management in developing countries

It is clear from the experience of the developed countries that environmental management is not a politically neutral, scientific activity. The need for environmental management emerges from the contradictions of economic growth in industrial society. The ineffectiveness of environmental interventions is closely related to the power of established interests in this society. The reliance placed upon techniques for containing environmental damage, rather than more radical intervention, reflects the fact that techniques are easier to research and evaluate than programmes or policies (Blaikie 1985, 40).

The situation in most developing countries is even less conducive to effective environmental interventions. This is partly because of the emphasis placed on project planning:

> Projects are planned in a vacuum created by the death of the development planning dialogue. The emphasis on project planning to the exclusion of development planning has been further accentuated by two mutually reinforcing conditions. First, development is ruled by the forces of the international market-place. Development planning cannot take place except within the conditions laid down by international capital markets (Hosier *et al.* 1982, 182).

Hosier and his colleagues go on to argue that the methods employed in project planning, particularly cost–benefit analysis, are essentially means to minimize the risk to capital. Despite the inclusion of social and environmental criteria in cost–benefit analysis, they rarely carry the same weight as financial criteria in determining the fate of a project. Looking at the decision-making inside a development agency Conlin (1985) makes a similar point: financial criteria are not only determinant in the project's inception, the disbursement of funds is considered of such importance that it is usually considered the principal responsibility of development agencies.

In the light of these observations it is useful to set out the differences between 'environmental planning' in the narrow, technical sense alluded to above, and 'development planning' as a wider exercise in designing more appropriate development policies. Table 7.3 draws liberally on the excellent work undertaken in Santiago, Chile by the United Nations Environment Programme (UNEP) and the Economic Commission for Latin America (ECLA). The table illustrates how each methodological characteristic of the planning

Table 7.3 Environmental planning methodologies in LDCs

Methodological characteristic	'Environmental' planning	'Developmental/ environmental' planning
Focus	Descriptive: what happens	Explanatory: why it happens
Variables	Intrinsic to the natural environment – soils, bio-mass, precipitation, etc.	Variables generated by environment: productivity, stability, carrying capacity, equity
Analysis	(a) Linear: cause–effect	(a) Linear and non-linear
	(b) Impact assessment	(b) Spatial and temporal discontinuities
	(c) Environment restricts or determines human activity	(c) Environment as a system that responds to human activity
Evaluation	(a) Existing pressures and resources only	(a) Different options given different resource uses
	(b) Short/medium time horizon	(b) Long time horizon
Policy instruments	(a) Direct: prohibitions, zoning, etc.	(a) Direct and indirect: modifying policy context
	(b) Resource specific	(b) Systemic – minimization of environmental risks and losses
Objectives	(a) Reduce negative impacts on environment	(a) Design environmental strategies to reduce risks and losses to system
	(b) Positivist – 'value free'	(b) Strategies linked to normative 'development styles'

Source: adapted liberally from ECLA/UNEP 1985

process is conceived differently: the variables under consideration, the kind of analysis undertaken, the breadth of options and time horizons and the policy instruments on which emphasis is placed. The broader conception of development/environmental planning is more holistic, has more heuristic power and considers alternative options and time horizons. The emphasis in this more integrative approach is on reducing the need for 'reactive' environmental planning, by seeking to remove the most environmentally harmful effects of 'development'.

The limitations of conventional environmental management are clear from the activities of the World Bank. Between 1971 and 1978 the World Bank made a total of 1342 loans and credits. In a majority of cases (845) there were 'no apparent or potential environmental problems' (World Bank 1979). Of those projects in which an environmental component was identified less than one-third justified calling on outside consultants. In the majority of cases, even when an environmental problem was identified, the Bank was content to rely upon their own staff to improve the environmental design and operation of the project.

This preference for 'in house' assessment took place against the backcloth of increased Bank lending for projects in environmentally fragile tropical areas. Although the World Bank had begun to incorporate environmental protection into its work, by 1970 the sums of money spent on environmental activities were derisory. In the case of most Bank-supported projects less than 3 per cent of the project costs were allocated to environmental protection, and most of this funding was for 'precautionary measures (which) were added when the project design was already well advanced' (World Bank 1979, 12).

This very limited funding was being made at a time when the assimilative capacity of ecological systems was being placed under increasing strain. In 1983 the four multilateral development banks loaned US$20 billion to developing countries towards a total project cost of three times this figure. Two areas which were particularly important from an environmental standpoint received enormous funding from these banks in the 1970s and early 1980s. These were projects to introduce more capital-intensive agriculture or ranching to tropical forest areas and resettlement schemes in the tropics. Between 1970 and 1977 the World Bank and the Inter American Development Bank together loaned US$5 billion to Latin American countries for livestock development. In Rich's words: 'No single commodity in the Third World has *ever* received such extraordinary outside support as livestock in Latin America' (Rich 1985). Similarly, resettlement schemes, which from both a social and environmental point of view were even more dubious, attracted enormous support from these banks. The World Bank authorized US$350 million for transmigration in Indonesia, moving hundreds of thousands of people from Java to Borneo and Sumatra. Brazil's development programme for the north west (POLONOROESTE) received US$443 million in 1981. It has been estimated that between 1979 and 1983 almost half a million people had been involuntarily resettled in projects approved by the World Bank (Rich 1985).

In view of the scale of multilateral-development-bank involvement, it is perhaps surprising that these institutions have paid so little attention to environmental factors in tropical development. Among the six thousand employees of the World Bank only one is trained as an ecologist. By May 1984 the Bank was rethinking its approach to the environment and adopting new environmental procedures. However, environmental and resource management were only incorporated in detail into the Bank's forestry paper and fared less well in other parts of the Bank's operations. In her account of working at the World Bank Catherine Watson (1985) lays stress on the fact that, although environmental guidelines were established for agricultural projects, it was not essential that they were followed. In most cases agricultural projects omitted to mention environmental problems and the small staff of the Bank's Office of Environmental Affairs had to mount their own investigation of projects. Watson estimates that only one in twelve projects received detailed scrutiny from her office. Her conclusion was that environmental proposals were only supported when they involved little financial cost. Some issues, like that of protection for tribal peoples, became Bank protocol because of the pressure mounted by the environmentalists, and because they did not require large expenditures. When environmental considerations threatened the economic viability of a project, they were dismissed as unrealistic. The primary consideration governing World Bank action was the need to increase foreign exchange in developing countries; hence the support given to ranching and export crops in the projects the Bank supported.

The practice of environmental management in Mexico

To appreciate the limitations of environmental management in a developing country it is useful to consider a specific case, that of Mexico. Although severe environmental problems have existed in Mexico for centuries, as we saw in chapter 6, they were not usually perceived as environmental. In 1972 the Secretaria de Mejoramiento del Ambiente (SMA) was established with a remit to report on environmental problems. In practice, during the presidency of Echevarria (1970–6) environmental protection was reduced to two concerns: pollution and public health. According to Godau (1985) this held advantages for the Mexican state in that nobody could oppose better health and the pollution issue enabled the state to make a pact with private interests. Within the state bureaucracy the organizations responsible for environmental monitoring, such as the SMA, were the weakest links in the public bureaucracy. Public environmental policy in Mexico was dedicated to denouncing the damage which underdevelopment inflicted on the environment (Godau 1985, 59).

Powerful state agencies, such as the Agriculture and Water Resources Ministry (SARH) or the Mexican government's petroleum agency (PEMEX) failed to co-operate with the environmental monitoring process. Air and water pollution were never subjected to serious criticism. Between 1976 and 1982 Mexico's oil reserves enabled the country to enjoy a brief expenditure boom and, in deference to this economic prosperity, environmental policy was switched from an industrial to a domestic focus. The problem was no longer air or water pollution, despite the deterioration in both; the problem was human waste and public health. Meanwhile the statistics published by SMA on levels of air pollution were unconvincing, even to the agency's own officials. By 1982 the deteriorating situation, particularly in the Federal District, persuaded the incoming president, de la Madrid, to establish an environmental agency with wider powers, the Office for Urban Development and Ecology (SEDUE). In view of the magnitude of Mexico's debt crisis, both the foreign debt and the domestic public sector debt, the only option open to the government was to legalize environmental pollution by acknowledging that the resources did not exist to deal with the problem.

Table 7.4 illustrates the problems confronted by environmental policy in Mexico. In the first column the various national plans and programmes with environmental content are listed. In the second column the environmental components of these plans are specified. In the third column is a list of the problems associated with attempts to implement these programmes. It is clear that, although the provisions for environmental intervention exist, their implementation is very uneven. Different plans carry different implications for the environment: in some cases reducing the 'externality' effects of economic growth; in other cases amenity or conservation goals are specified. The effect of pursuing environmental objectives within agencies dedicated to non-environmental ends is therefore often contradictory. These contradictions are present both within sectorally defined planning agencies and between them. The plan to develop agriculture, for example, seeks to promote continuing

Table 7.4 Environmental management in Mexico 1984

Plans and develop-ment programmes	Environmental components of plan	Implementation
1 Global Development Plan	reduce pollution	(a) Different plans often have contradictory implications for the environment, e.g., the agricultural development plan (2) is geared to incremental growth, the energy (5) and tourist plans (3) imply reductions in growth.
2 National Programme for Agriculture and Forestry	conserve renewable natural resources	(b) Public administration of development plans is organized sectorally. Problems and solutions are inter-sectoral.
3 National Plan for Tourism	maintain nature for human access	(c) No specific provision is made for environmental programmes within public expenditure budgets.
4 National Plan for Urban Development	to develop natural resources for human settlements	(d) The strategy and instruments for implementing the various environmental objectives remain undefined. No provision for environmental impact studies.

Table 7.4 Environmental management in Mexico 1984 (*continued*)

Plans and develop-ment programmes	Environmental components of plan	Implementation
5 Energy Programme	to protect the environment from energy growth (esp. petro-chemicals/	(e) Limited fiscal measures against polluters and few efforts to reduce pollution.
6 Urban Develop-ment for Federal District (DF)	to reduce urban pollution	(f) 'Biosphere reserves' estab-lished by CONACYT in Durango, Jalisco, Quintana Roo and Sonora, incorpor-ate local populations in con-servation activities. These show great potential.
7 Others e.g. plans for agro-industry, fisheries industry, education, co-operatives, housing, science and technology		

Source: Information from Alejandro Toledo (1985), *Como Destruir el Paraiso*, Mexico City, Centro de Ecodesarrollo.

agricultural growth *and* the conservation of renewable natural resources.

Many environmental problems are intersectoral in nature, and responsibility for them is shared between different ministries and government departments. Environmental programmes are also underfunded, undefined and scarcely ever evaluated. Where measures exist in law to prevent environmental damage, the agencies whose responsibility it is to ensure enforcement usually lack professionally qualified people and political muscle. There is then neither the expertise nor the political backing for decisive action on environ-mental degradation. The benefits to be derived from implementing environmental measures are often intangible, while the financial advantages which public sector employees gain from powerful economic interests opposed to environmental measures, are very real indeed.

Defining an alternative: indigenous environmental management

It would be inaccurate to assume that environmental management is the sole prerogative of technically trained people from developed countries. Many rural people using simple technologies in developing countries possess a fund of information about their environment and can effectively manage that environment in ways that are sustainable in the long term. However, the experience and knowledge of such people is only rarely incorporated in the formal structure of rural environmental planning in developing countries. In the previous section some of the limitations of the kind of environmental management with which we are most familiar in the North were identified. It is now necessary to ask what are the implications of seeking to learn from indigenous experiences of managing the environment for the development of rural societies in the South?

First, it is clear that among indigenous people (tribal or 'native' peoples) of developing countries, sustainable practices are adhered to because traditionally they were the only guarantee of survival. It is not surprising then that native peoples should regard their role as one of stewardship, particularly in societies where the use of natural resources was not necessarily tied to their ownership. Ovington and Fox (1980, 58) tell us that:

> In the case of the Australian Aboriginals, identification (with their environment) after two or three thousand human generations is so complete that they do not separate themselves from their environment. They see themselves as part of the landscape, not apart from it . . . Once people are taken out of close contact with the natural environment which sustains them, the perception of dependence fades. The notion that the ultimate source of food, shelter, new crops, new drugs and new materials must ultimately come from the Earth loses force in this change of perception.

Not all human groups live as closely, or as symbiotically with their environment as the Australian Aboriginals, but the recognition that people depend upon their environment for their survival is widespread. A particular cultural group will not necessarily respect the constraints on resource use stemming from the *theoretical* carrying capacity of land, but the knowledge gained from sustainable-resource use forms part of the environmental *practices* of most indigenous populations. People are used to interfering and modifying their own management practices and have grown used to living with the consequences of their actions. It is much more difficult to predict, or justify, interference in culturally rooted practices in the interest of management strategies that are imposed from outside. As Nowicki (1985, 285) puts it: '. . . there is a difference in making a punctual investment which increases the overall productive capacity of an ecological system which supports itself, and the investment which changes initial circumstances so that the system is no longer self-supporting'.

It should be remembered that, within most indigenous groups whose livelihoods are not dependent on commodity production for the market, there is no internal tendency towards the maximization of profits or the creation of an economic surplus. The 'balancing act', as argued in chapter 2, is frequently between population dynamics and natural resources. The more intensive use of traditional techniques is linked to a strategy for minimizing risks and widening options in the face of the insistent (and usually indirect) process of capital accumulation. In the course of development indigenous environmental knowledge is often lost, because it becomes less relevant to the new situation and because it is systematically devalued by the process of specialization around competitive production for the market.

Traditional environmental knowledge is not only devalued by development institutions, it is also largely overlooked in the environmental management literature, as Norgaard (1985a) acknowledges. This is partly because of the way such knowledge is recorded in the cultures of native peoples. Without knowledge of the culture a people possess one is unlikely to be aware of their knowledge of their environment. The corollary is that, if we want to know how ecological practices can be designed which are more compatible with social systems, we need to embrace the epistemologies of indigenous people, including their ways of organizing their knowledge of their environment. However, as Norgaard argues, traditional knowledge is location specific and only arrived at 'through a unique co-evolution between specific social and ecological systems' (1985, 876). This knowledge is not easy to incorporate into 'scientific' knowledge since experiential learning requires an evolutionary rationale, and one which is different from that of bureaucratically managed institutions. These differences between what Sohn-Rethel (1986) calls 'societies of production' (including indigenous cultures) and

'societies of appropriation' (including modern development institutions) is important to acknowledge. In most traditional societies
outsiders can frequently appreciate that practices make sense, but
the epistemology employed in arriving at these practices is usually
obscure to outsiders. It is culturally coded in ways that even
anthropologists find difficult to translate. On the other hand,
societies in which 'rational' scientific knowledge is routinely
employed, like our own, are also ones in which social practice is
frequently irrational. Examples are not difficult to find. Under the
guise of technological policy we have been led from a situation in
which governments seek military supremacy towards the creation of
a technology (nuclear energy and nuclear weapons) which puts our
very survival in jeopardy. Our control over our environment does
not match our knowledge of our environment. These different
epistemological positions are illustrated in table 7.5 below.

Table 7.5 Environmental knowledge and commodity production

		Practice	Theory
(a) Societies of production	Rational	X	
	Irrational		X
(b) Societies of appropriation	Rational		X
	Irrational	X	

In societies of production (a) practice is rational in that it makes sense to outsiders,
whereas the epistemology employed is often impossible for outsiders to understand.

In societies of appropriation (b) social practice is irrational in that people have lost
control over their environment, but scientific knowledge replaces traditional
epistemology.

Source: adapted from Sohn-Rethel 1986

Clearly indigenous people 'see' nature differently, precisely
because their practices acknowledge its centrality. Studies in
ethnobotany reveal that people in societies characterized by simple
technologies are aware of differences in nature which are invisible to
specialists from outside. Dandler and Sage (1985), for example,
report that in one Aymara community in the Bolivian Altiplano five

peasant households named 38 'sweet' and 9 'bitter' varieties of potato that they themselves cultivate. They are aware also that each variety possesses advantages and disadvantages as part of a strategy for minimizing risks. Their interest in maintaining crop diversity is based not on a belief in diversity for its own sake, but on the knowledge that diversity reduces their environmental vulnerability.

Table 7.6 expresses the relationship between indigenous knowledge and the survival strategies of rural people who have yet to become separated from their environment by market processes.

Table 7.6 Indigenous knowledge and survival strategies

1   Production of use values ('subsistence')
2   Ecological adaptation and sustainable management
3   Indigenous technical knowledge (ethnoscience)

micro-level practices	*protect against*
ecological diversity	climate
crop diversity	hazards (pests)
varietal diversity	genetic erosion

Minimizing uncertainty strategies
nutrient conservation
sustainability
nutritional complementarity
labour spreading

Counter-seasonality strategies	
habitat seasonality	climate/environment
seasonal adversities	nutrition, health, labour, natality etc

Social strategies
supra-household labour obligations
community owned resources

Source: adapted from Sage 1985.

Their knowledge is based on the production of use values and the adaptation of their agricultural practices to ecological conditions. Indigenous technical knowledge informs these practices, in conserving energy, in cultivating crops, in combination with animal rearing, and in other farming/conservation techniques.

The use of indigenous knowledge is linked to the strategies which the culture has devised for coping with risks. These micro-level practices protect against vicissitudes in climate, attack from pests and genetic erosion, through maintaining diversity, in the ecological system, in crops and in genetic materials. The components of these strategies allow a sustainable system to be reproduced in which biological nutrients are conserved, and food consumption meets different nutritional needs. At the same time the labour effort expended in production and consumption is spread over the agricultural calendar. Another important factor is that these strategies reduce fluctuations due to seasonality, both in the habitat and in seasonal vulnerability to disease. Finally, organized around the household or the community, social strategies exist which ensure that access to commonly held resources is linked to the investment which a household makes in the community. The strategies adopted do not necessarily succeed in ensuring adequate livelihoods, but they are designed to reduce the risks to those livelihoods.

As Denevan *et al.* (1982) have shown, strategies like swidden fallows, which involve a combination of annual crops, perennial tree crops and natural forest regrowth, constitute a sustainable management system for the Bora Indians of the Peruvian Amazon. Under this system of shifting cultivation swiddens are almost never completely abandoned but are held in a transitional stage as their productivity declines. These practices reduce the ecological and social vulnerability of the Bora and lend themselves to adaptation. In effect they are systems of agroforestry, designed and managed by the indigenous people themselves. They provide the guarantee that both food crops and perennial cash crops will continue to be cultivated without doing irreparable harm to the environment.

The knowledge that, under more densely settled populations in the past (Denevan *et al.* 1976), large areas of Amazon forest may actually have been stages in productive swidden fallows leads Hecht, Anderson and May (1985) to talk of 'the subsidy from Nature' in the humid tropics. They point out that extractive activities have rarely been given importance in tropical areas of developing countries, when these activities were in the hands of local, native peoples. The small scale of the extraction and their role in meeting use values meant that these activities were looked upon as evidence of the poverty and poor husbandry of indigenous groups. Hecht and her colleagues demonstrate that small-scale extractive activities are an important source of income, particularly for women, who are usually denied access to other ways of earning cash income. They argue that the main thrust of rural development has been 'to strengthen the ties between agriculture and markets with an emphasis upon technological change and the production process (while) . . . the most sustainable and possibly most efficient land uses incorporate natural biotic communities as well as domesticated crops and animals' (1985, 2).

The 'subsidy from Nature' therefore takes two forms: it refers to the forest biomass which accumulates during fallow periods and forms the basis of successional systems of cultivation, and it also refers to the existence of subsistence products on a small, but important, scale. These elements of the livelihood of people are both invisible to most rural development planners and do not enter into their calculations of an area's development potential. In this respect the relationship with nature involves people in spheres of activity that remain obscured from the vision of environmental managers. Like women's involvement in non-market exchanges, the domestic production of use values and reproduction (both biological and social) constitute spheres of environmental activity that form a much larger part of household livelihoods than is usually acknowledged. The low visibility of these activities is closely related to the fact that they are so often performed by women and children.

To appreciate the potential gains in widening our view of environmental management we need to be aware not only of differing epistemological positions, but even of different cosmologies. Incorporating the environment into development planning implies incorporating cultures and cultural perspectives. The problem of how to manage the environment effectively in developing countries is usually raised in terms of 'their' cultural adaptation to development processes. Perhaps the more urgent question is whether 'we' are prepared for the cultural adaptation that is required of us. The reference to cosmologies was deliberate, since the view that people take of their environment is intimately linked to their conception of their own place in space and time. A recent collection edited by Jeffrey McNeely and David Pitt (1985) provides evidence that people in small-scale societies, the tribal or indigenous people who make up 200 million of the world's population, often view the environment within a long-term time horizon, like many ecologists. In pastoral societies, for example, significant events may be separated by decades. The inability to appreciate this fact helps explain why, when people judge environmental losses to have become critical, development personnel are often the last to know. It is essential to know when environmental problems are unsustainable, to people as well as in technical terms, and the answers lie partly in cultural interpretations of 'crisis'.

The cultural categories that people use to classify and understand their environment should be important to all those interested in its management: first, because local people possess close acquaintance with their environment over a long period; second, because the more we discover about the components of the environment, the more we are forced to consider their interrelationship. A holistic concern with the relationships between components of human population and the environment, a discovery of post-industrial society, is commonplace in most traditional cultures. We need only consider, for example, our enthusiasm in discovering that what we take from nature to create a balanced diet itself affects the balance achieved in nature. These ideas are beginning to attract attention as a way of escaping from, or minimizing the pressures of our society, but among the tribal populations of developing countries they are almost cultural edicts. Finally, it is necessary to attach weight to cultural definitions of the environment because environmental planning cannot work without the participation of people, and this participation depends on cultural understanding and mutual respect.

Undertaking research in Sierra Leone, Paul Richards (1985) sought advice on weed growth from three groups of people: university botany and geography students, farmers and extensionists. He found that the agricultural extension staff shared similar categories with those of the university students, but both were quite unlike the categories employed by farmers. He also found that farmers' units of land under crops usually represented labour inputs, not land areas. Farmers used categories such as 'the amount of work that could be undertaken in a standing position', and rows of crops corresponded to these labour categories. Similarly farmers explained why agronomic practices did not work by making causal links between plants, animals and humans, but they were not the links made by professional staff. As most farming-systems research has indicated, farming systems are essentially experimental laboratories for farmers in which they are prepared to try new methods, provided the risks are not unacceptably high. The lessons need to be learned by the scientists. What research agronomists call 'recommendation domains', or local environments for which recommendations can be made, often do not correspond with the categories used by farmers. Research undertaken by the International Potato Centre (CIP) in the Mantaro Valley, Peru, helped identify ecological zones for potatoes that were different from the technically designated zones (Horton, 1984). Among the research findings was that the *ticpa* system of potato cultivation, employing native potato varieties, no tillage before planting, hand power using a plough and very little chemical fertilizer and pesticides, produced higher net returns than the system employing modern varieties, tractor power and high levels of chemical inputs. The research team concluded that:

> These empirical findings were in sharp contradiction to the assumptions of many CIP scientists and development experts working in the Andes. They helped to destroy the myth that traditionalism among the small-scale farmers is a major barrier to the transfer of technology (Horton 1984, 42).

It would be an exaggeration to claim that farmers need to do the 'managing' and we need to do the 'learning' about the environment, since we need both sets of skills and their successful use requires an epistemology and a practice. Nevertheless, it is difficult to exaggerate the gains from seeking to reverse the process under which 'we' do all the managing, while 'they' operate within the space, and with the technology, that we provide. Figure 7.2 illustrates the redirection that is necessary if we are to move from present-day 'environmental managerialism' to a more collaborative view of environmental management which takes its cues from the environmental users rather than the outside 'experts'. To achieve a more efficient and accountable system of environmental management we should take counsel from Leff (1985, 264–7):

> the traditional practices of pre-capitalist societies can serve as a starting point for the implementation of more efficient modes of management under the principle of preserving its basic ecological structures and the cultural integrity of the people . . . research in the fields of ethnolinguistics and ethnotechnology will serve to rescue some traditional practices and the means of production of a culture, and to assess the ecological and historical viability of different modes of management of the natural resources.

Figure 7.2

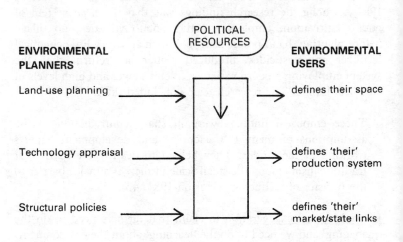

ENVIRONMENTAL 'MANAGERIALISM'

ENVIRONMENTAL PLANNERS		ENVIRONMENTAL USERS
Land-use planning	→	defines their space
Technology appraisal	→	defines 'their' production system
Structural policies	→	defines 'their' market/state links

POLITICAL RESOURCES

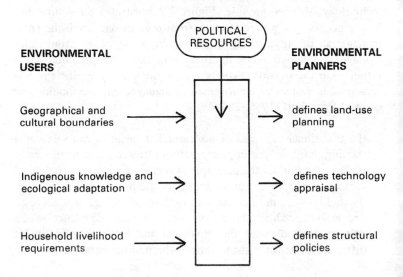

COLLABORATIVE ENVIRONMENTAL MANAGEMENT

ENVIRONMENTAL USERS		ENVIRONMENTAL PLANNERS
Geographical and cultural boundaries	→	defines land-use planning
Indigenous knowledge and ecological adaptation	→	defines technology appraisal
Household livelihood requirements	→	defines structural policies

POLITICAL RESOURCES

Environmental movements in the South

It is sometimes argued that social movements in developing countries are unlikely to embrace environmental demands, since the contradictions of the development process leave poor people with little option but to make ever increasing demands on their resource base. Even in relation to the urban environment such a line of argument can be challenged. The growth of new urban social movements in Latin America, for example, clearly owes something to environmental consciousness (Slater 1985). The role of ecological considerations in rural social movements in the South is certainly ambiguous but it deserves to be examined. It is important that we do not reserve the term 'environmental movement' for social protests created in the image of similar movements in the North. The two principal components of environmental movements in the South are of marginal importance to most movements in the developed countries. They are that those who constitute the 'movement' are engaged in a livelihood struggle and, secondly, that they recognize that this livelihood struggle can be successful only if the environment is managed in a sustainable way. As nature is transformed under capitalist development and 'natural' resources are created, social struggles are initiated which resist the incorporation of nature in wider spheres of accumulation. The concern of these movements with the distribution of resources is usually linked to ideas about the alternative uses to which they could be put. The struggle to create the conditions of existence necessary for social and biological reproduction, outside the spheres of capitalist accumulation and market-oriented resource use, is a struggle to effect profound environmental objectives. Movements to realize these objectives are necessarily environmental movements. In this chapter three such movements are examined: the Chipko movement in India, the Green Belt movement in Kenya and recent ecological movements in Mexico.

The Chipko movement in India

The popular interpretation of India's Chipko movement is that women have acted spontaneously since the 1970s to protect trees from being felled. However, as Shiva and Bandyopadhyay (1986) show, this misrepresents the movement's history and objectives.

They point out that since the last century the state has continually encroached upon the rights and privileges of people to forest resources. The resistance to this encroachment has taken traditional Gandhian form, in the power of *satyagraha*, or peaceful non-co-operation. In the forest areas of the Gharwhal Himalaya the style of protest which had originally been directed at the British, for attempting to sell off community forests, was revived and used against the Indian state. The movement which began in the Gharwhal highlands has now spread to much of upland India. In its revived form the civil disobedience practised by Chipko adherents has taken on an increasingly ecological character: 'Although it had its roots in a movement based on the politics of the distribution of the benefits of resources, it soon became an ecological movement rooted in the politics of the distribution of ecological costs' (Shiva and Bandyo-padhyay 1986, 1).

The history of struggle for control of India's forests is an illuminating one. Before the full impact of colonialism, conservation strategies appeared to play a large part in the lives of Indian hill people. Forests were managed as common resources with strict enforcement of informally agreed codes of management. Large tracts of natural forest were maintained through this careful husbandry, and the selection of tree species was a recognized part of village forest and woodlot conservation.

The colonial impact on forest resources seriously undermined these strategies. First, changes in land tenure, such as the introduction of the *zamindari* system, transformed commonly held village resources into the property of private landlords, and village resources were destroyed as a result. The population whose livelihood was most affected had increasing recourse to natural forests in an attempt to survive the transition. Second, there was large-scale felling of these natural forests as the British administrators met the increasing demands of the shipbuilding industry and the expanding Indian railway network. The process of deforestation was soon so advanced that the British colonial administration sought to ensure its supply of timber by reserving forest areas for commercial extraction. The conservation of forests, under colonial rule, was directed at the maintenance of forest revenues rather than environmental objectives. This involved refusing local people their traditional rights to use forest resources. It also led to unsustainable practices in the reserved areas, where the stability of forest ecosystems was destroyed and ecologically unsound practices were introduced.

During the early 1930s the movement of resistance to forest enclosure spread throughout India. This movement was concerned with resistance to the transformation of the forests into a commodity which imperilled the unity of people and nature, a hallmark of resistance to colonial rule. The forest satyagrahas were especially successful in regions like the Himalaya, the Western Ghats and the Central Indian hills. The non-violent protests of Indian people were suppressed by the British rulers and unarmed villagers were killed. The transfer of power to an independent Indian government did not change the way the forests were managed. On the contrary, subsequent Indian governments became dedicated to the same principles of forest management as the British, in the interest of achieving high rates of economic growth.

According to Vandana Shiva, a participant in the Chipko movement as well as a formidable academic critic, the Gandhian world-view, which laid emphasis on justice and ecological stability, formed part of the consciousness of women in the hill areas of Uttar Pradesh. Sunderlal Bahuguna is one of the leaders, prominent in today's Chipko movement, who acknowledges this debt to Gandhian philosophy. Today, at 60 years of age, he is engaged in strengthening the philosophical base of the Chipko movement and its Gandhian view of nature. At the same time the rapid spread of resistance in Uttar Pradesh and its success in changing forest management practices are partly due to the awareness created by folk poets like Chanshyam Raturi, as well as a legion of grass-roots organizers.

The basis for a revived environmental movement existed, as we have seen, before the ecological instability of the Himalayan forests was fully recognized. In its early stage this movement attempted to stop the auctioning of forests for felling by contractors. In an attempt to mobilize support for the movement, Chanshyam Raturi wrote songs of popular protest, reminding the hill people of the need for forest protection. By 1974 the now famous Chipko movement was born.

Chipko, like the earlier forest satyagrahas, is aimed at integrating the conservation of forest resources with maintenance of livelihoods and the preservation of culture. Not surprisingly it is women who are the main bearers of this tradition. The first Chipko action took place in April 1973 when a group of villagers demonstrated against felling ash trees in the Mandal forest. Almost a year later a large group of women saved another part of the forest from the contractor's axe. Soon afterwards the government stopped the contract system of felling in Uttar Pradesh, but the influence of Chipko resistance spread throughout the area. In June 1977 a meeting was planned of all activists in the hill areas of Uttar Pradesh state, which served to widen the objectives of the movement in opposition to resin-tapping as well as deforestation.

During 1977 and 1978 several protest demonstrations were successfully held in the Gharhwal Himalaya. On one occasion the large number of women guarded the forests from contractors; on another occasion a fast was undertaken in the forest itself. Groups of women began to read from ancient texts about the value of the forest and the ecological slogan was born, 'What do the forests bear? Soil, Water and Pure Air!' The movement placed emphasis on forests as areas of soil and water conservation, rather than as sources of timber and resin. The demand of the Chipko movement was not for a bigger share in the commercial development of the forest for local people, but the design of an alternative system of environmental management in which fuel, food, animal fodder and recycled organic waste all played a part. Today the Chipko movement is divided. The first group consists of those who argue that ecological costs can be met only by satisfying primary human needs. The second group is made up of those who argue that relocating manufacturing activities in the hill areas will reduce poverty, provided that the raw material base for these activities is developed. To some extent this division may anticipate divisions in the wider ecological movement, as development institutions incorporate some of the demands of grass-roots environmental movements, while continuing to support productivist programmes of resource utilization.

The Green Belt movement in Kenya

Support for mass tree planting by Kenyan women began in June 1977 on the occasion of World Environment Day. The activity was co-ordinated by the National Council of Kenyan Women and developed into a programme 'that approached the issues of development holistically and endeavoured to build on local expertise and capabilities' (Maathai 1986, 20). One of the leading figures in the movement, Wangari Maathai, has recently described the genesis of the movement in the action which the Women's Committee took together with a small group of very poor rural women (Maathai 1985, 1986). Unlike the Chipko movement in India, the initial steps towards forming a movement were taken by Kenyan women outside the communities; they identified forest losses as among the most serious causes of soil erosion and land degradation in Kenya. The Green Belt movement has grown slowly but surely around a programme of reforestation, environmental awareness and better livestock and crop production. Although women are the principal activists, men also participate. In an interview with British feminists Wangari Maathai (1985, 16) said:

> one of the things I like about the movement (is) that here is an activity initiated by women, but men participate in it because it's a development issue. This has also helped the women's cause because it's the women who provide the leadership.

Once contact has been made with community members who wish to participate in the movement, they are urged to apply to headquarters indicating whether they wish to participate by growing tree seedlings or by planting them. The participants then carry out certain work before they can receive trees, preparing the ground for tree nurseries or digging holes for planting. This work is checked by the Green Belt movement field staff.

Participants in the programme then attend meetings to teach them forestry techniques, while field staff monitor their project. While seedlings are being produced the community is being persuaded to plant trees. When the seedlings are ready they are distributed to members of the community. In this way over 920 public green belts, each with over one thousand trees, have been planted and another 15,000 private green belts, under the management of small farmers, are registered with the movement.

Since about 90 per cent of Kenya's 19 million people live in the rural areas and most of them use fuelwood, the demand for timber is unlikely to decline. Similarly in urban areas wood is converted to charcoal for domestic use. In many parts of the country soil fertility has fallen and famine has become a recurrent phenomenon. This loss of soil fertility is connected with the indiscriminate felling of trees and the gradual encroachment on forests by human and livestock populations. Forests cover about 2 per cent of the land formerly forested, and over two-thirds of Kenya's land area is arid, semi-arid or desert land. Maathai argues that tea and coffee plantations should not be developed while so much lost land has still not been reclaimed (1985, 16). The task of building a sustainable development path is made more difficult by the legacy of colonial exploitation, under which most of the best land in Kenya was reserved for the white population, while black Kenyans were forced on to increasingly marginal land.

Mexico's environmental movement

Mexico has two environmental movements: the first movement is composed of mainly middle-class, urban people and bears some resemblance to the kinds of environmental movement in developed countries with which readers of this book will be familiar. The second movement is closer to the movements in other developing countries described above. It is made up of poor people, both urban and rural, whose attempts to improve their livelihoods are linked, increasingly, to sustainable practices. In the Mexican case this latter group receives considerable attention from radical scholars and policy makers, who share with the poor the conviction that 'the term development now appears mainly in jokes' (Esteva 1986, 15). The problem for the environmental movement in Mexico is how to integrate two perspectives, one of which is the outcome of specific livelihood struggles, while the other is a consequence of greater environmental awareness among a small fraction of the educated élite.

Let us begin by considering first, the perspective of Mexico's environmentally conscious middle classes.

The conservation movement has been slow to develop in Mexico. Most of the 20 or so conservation groups that exist operate at a local level with very small memberships. The exception is the Mexican Ecological Movement (MEM) founded in 1982 by Alfonso Cipres Villarreal, who for the last 15 years has been an indefatigable and often isolated critic of environmental neglect. Today the MEM has 52,000 members organized around local groups or 'clubs'. Membership is drawn almost entirely from the middle classes, who are also seen as the most fertile ground for the movement's future growth.

Its narrow membership notwithstanding, the MEM has proved

surprisingly effective as a pressure group. The organization sees itself as a necessary corrective to the often sanguine public pronouncements of the Mexican government. It has resisted attempts by the governing party (PRI) to provide funds in return for political control and it has constantly rejected the offer of funds from corporate industry, including, after the Bhopal tragedy, Union Carbide. The MEM is highly critical of Mexican government policy, while emphasizing its allegiance to the constitution, and the concept of 'Revolutionary Nationalism' in Mexico. On the several occasions when its leaders have been received by the president (the last in November 1984), the MEM has been at pains to distinguish between its anti-government stance and its unwillingness to act as a party of opposition. Cipres Villarreal argues that the MEM will become a political party only when it has two million members and realizes that this is not a feasible objective today.

The principal method adopted by the MEM is to win the vocal support of prominent Mexicans in both the arts and the sciences. Although public demonstrations have been held, some of them large, direct action is condemned. Moreover, there have been some notable successes in seeking out members from within the technical and cultural élite. Writers like Octavio Paz, Juan Rulfo and Jose Luis Cuevas have all joined the movement and given it public support, as have artists like Tamayo and Ignacio Beteta. Scientists, such as the Nobel Peace Prize winner, Norman Borlaug, are also members of the movement.

In the view of the MEM the Mexican state seeks to assume responsibility for resolving ecological problems because it is afraid to put its faith in civil society. Had the Mexican government of De la Madrid sought genuinely to adopt radical proposals, it would have established an autonomous institute to act as its watchdog, instead of another bureaucratic organization to dispense favours and maintain social control. The MEM sees the new Environment Ministry, established under President de la Madrid, as a device by which the PRI (Institutional Revolutionary Party) seeks to pre-empt, and ultimately control, environmental organizations within civil society. This, they feel, is the Mexican 'way', but not the correct way, to proceed.

The MEM's ideology is more radical than might be supposed from its membership. There appears to be a genuine interest in appropriate technology, alternative health care and housing policies.

The determination exists to raise questions about the Mexican state's public pronouncements and activities 'so that the people do not lose sovereignty over ecological issues'. As their literature puts it, 'the only democratic right that the Mexican people possess is the right to pollution'. Several attempts have been made to infiltrate the MEM, notably by the Roman Catholic Church, which in Mexico lacks a social and political base compatible with its widespread but unofficial following. Unlike the situation in some West European countries, the Roman Catholic Church in Mexico is not looked upon as progressive by activists in the environmental movement. The major and right-wing party, the Partido Autonoma Nacionalista (PAN) is uninterested in the environment as an issue because its finances are derived from big business which in Mexico is largely antithetical to environmentalism. The orthodox Left parties have never taken environmental issues seriously. They are divided and poorly organized.

Talking to MEM members one is struck by the absence of a clear social programme to which environmental issues could be linked. Unlike the European Green Movement, the influence of feminism and pacifism is not great, although the movement does pride itself on its 'utopianism' and is seriously worried by the risk that, as it grows in size, it will become more bureaucratic. The movement's ideology emphasizes what it terms a *vision indigena* (or indigenous vision) which it argues has been replaced historically by the *vision colonial* (colonial vision) in which Mexicans have been brought up. Unlike the conservation and environmental movements in Europe and North America, activists believe that radical action on the environment should take as its point of departure the need to break the knot of economic and cultural dependency which ties the South to the developed countries.

The second type of environmental movement in Mexico is different in social composition and political practice, although ideologically it shares common ground with organizations like the MEM. It is made up of poor people whose daily life and culture gives weight to sustainable development objectives. A network now exists of social scientists and professionals who seek to learn from the experience of popular groups. This consortium (ANADEGES or Analysis, Development and Self-Management) is a non-profit organization which seeks to build bridges with popular organizations, especially in rural areas, through projects such as agro-forestry and social forestry. In the view of ANADEGES there are only two options open to Mexico: either to follow the path set by international specialization, in which people lose control over their own lives, or to seek a more 'autonomous' and less authoritarian kind of development. In their view 'basic needs' are best defined by people themselves in the context of their own culture which requires a positive re-assessment of traditional ways of using the environment: *adobe* rather than brick houses, holistic medicine, *morada* (living space) rather than industrialized housing and transport.

Support for such ideas comes from internationally known thinkers like Ivan Illich, Andre Gorz and Rudolf Bahro, but the imprint is unmistakably Mexican. ANADEGES members are currently working with people in the urban *barrio* of Tepito, only eight blocks from Mexico City's main square, the Zocalo, where for over three centuries families linked by close community ties have combined self-provisioning, including keeping animals, with traditional trades and crafts. The experiences of the Tepito community are instructive. They illustrate the way in which crisis conditions can prompt collective environmental action.

Table 7.7 Environmental quality and alternative value systems: a Mexican perspective (ANADEGES)

Traditional 'autonomy'	Modern 'specialization'
1 *comida* (food and nourishment)	food provisioning
2 *salud* (health and well-being)	medical services
3 *morada* (dwelling space: intimacy and society)	housing and transport
4 *cultura* (human culture)	education and educational services

The left-hand column represents facets of an 'autonomous' and 'plural' system of values. The right-hand column represents the formal institutional and specialized version of these concepts in modern industrial society. In the view of ANADEGES Mexican society retains enough of the traditional value system to opt for the more autonomous road.

Tepito was one of the areas of the city most affected by the earthquake in September 1985 which claimed at least five thousand lives. The area contains approximately 120,000 people, most of whom live in overcrowded accommodation only eight blocks from the Zocalo, or central square. In Tepito many of the housing units, known as *vecindades*, collapsed from within following the earth-quake. The emergency relief co-ordinated under the Mexican government's antiquated DN-III plan did not arrive and, as in many other poor neighbourhoods, volunteers and local residents provided their own relief services. A local leader, Felipe Ehrenberg, informed the foreign press that, although the housing situation in Tepito was appalling, people were unwilling to accept housing outside the zone. They were asking for nothing short of the reconstruction of the same zone and feared that landlords would use the disaster as a pretext to force them from rent-controlled housing. Reports suggested that in another part of Tepito residents were organizing themselves around a community sports centre which they had turned into a temporary shelter and distribution point. The local government administration, the *Delegacion*, offered free labour and technical assistance to help rebuild the area, provided that the building materials were purchased by the tenant or owner. Tepito residents were not reassured by this since most of them are far too poor to be able to purchase such materials themselves.

In the days following the earthquake Tepito residents joined others in demonstrations, one of which (on 27 September 1985) involved over four thousand people, who marched to the presidential residence, Los Pinos, demanding emergency aid, water and housing. Accusations were made that corruption and lax enforcement of building codes had contributed to the disaster. This was particularly true at the large Tlatelolco apartment complex where there had been longstanding complaints about delayed repair work. Clearly some areas of the city were more susceptible than others to earthquake tremors. Central areas, like Tepito and Tlatelolco, were built on landfill spread over the soft lakebed soil that consisted of sand and volcanic ash. As we saw in chapter 6 before the Spaniards filled in this land, during the sixteenth century, the area that is today Mexico City had been a lake complex of raised-bed agriculture (chinampas) and adobe houses, with a population in excess of 300,000 people.

To understand the anxieties of Tepito's residents that they might be relocated elsewhere, we need to remember that where people live in Mexico is decided, for many people, by the government. In this case local people felt the government wanted to relocate them rather than face the public embarrassment of seeing people rebuild their own communities. Instead the Mexican government talked of the need for decentralization, taking its cues from the political culture of

Mexico City, where civil servants fear ostracism and isolation in provincial backwaters. Most of the residents of Tepito have nothing to gain from being relocated outside their neighbourhood.

Many of the families which settled in Tepito have long associations with the area. In some cases their ancestors arrived in the area during the eighteenth and nineteenth centuries. This was a barrio of small craftsmen, resembling the medieval guilds found in Europe. The legal framework prohibited the establishment of factories and offices. Instead, the people who settled in Tepito performed a service function to textile workers, public employees and other urban groups. They were petty-commodity producers in the classic sense referred to by Marx.

The social bonds between Tepito families included kinship and close 'godparent relations' of considerable importance in providing security for the poor in the area. The closest links in Tepito, however, were not between godfather and godsons, but between godfathers and fathers. Godparent ties are essential to the apprenticeship system, as boys learn trades that have flourished in the area for centuries. The families that live in Tepito are united by common origins, blood ties and godparent relations in ways that suggest rural community bonds rather than urban anonymity. No area of a Latin American city lends itself less to the description associated with Oscar Lewis: 'the Culture of Poverty'. Poverty exists, but Tepito's residents at least are representatives of one of the New World's oldest urban cultures.

Many of the households in Tepito have achieved high levels of self-sufficiency. They not only work *and* live in the same building, they also keep animals inside their *patios* (courtyards). Most of their clothes are also washed in these courtyards. A phrase has been coined to describe this lifestyle which marks out Tepito: the *patio-taller* or workshop/courtyard. In the eyes of ANADEGES it expresses the unity of purpose behind Tepito's population, a culture which is urban but pre-industrial in many respects. It represents an alternative to the chaos, alienation and environmental degradation of Mexico City.

Most areas of Mexico City do not resemble Tepito. Most of the 18 million inhabitants live in parts of the city quite unlike the inner-city area around Tepito. Almost half the built-up area consists of irregular settlements where land was either invaded, subdivided or sold, illegally, within the borders of 'agricultural' *ejidos* or peasant communities. Each of these areas has objectives for their own development, and an interest in their environment, which is often at odds with that of the politicians and planners.

Environmental movements have been discussed in this chapter in the context of wider social struggles to retain control over the natural environment often in the face of opposition from development agencies and governments. It is clear from the cases discussed here that there are two essential characteristics of environmental movements in developing countries: they are supported by people engaged in a livelihood struggle, and this struggle is linked to sustainable objectives. Where poor peoples' livelihoods are at stake, environmental movements will incorporate conservation objectives only within the context of basic needs. What makes the social movements of the urban and rural poor into environmental movements is that they seek to define the benefits of development in terms of basic environmental requirements: for energy, water, food and shelter. In the process such movements can become as interested in containing the costs of development as they are in enlarging the benefits.

This chapter has argued that environmental management must make use of social movements dedicated to environmental ends. It must also make use of the knowledge and experience which people possess about their environments. There are reasons for believing that sustainable development might one day be more than empty rhetoric if these issues are taken seriously. Before we assess the situation further, however, we need to recognize that the technical frontiers of sustainability are not fixed, and that what people can achieve for themselves is partly governed by their freedom of action within new technological limits. It is to this question that we turn now.

8

The frontiers of sustainability

Earlier chapters of this book have looked at how the environmental consequences of development have posed a contradiction for capitalism, and how that contradiction has been addressed. As the environment has become internationalized it has been transformed, and as it is transformed social struggles are mounted for the control and ownership of natural resources. The attempt to import solutions to environmental problems from developed countries, in the form of orthodox environmental management, therefore assumes more importance as the ecological crisis of development deepens. At the same time the potential exists to understand the historical dimension of environmental change by paying more than lip service to the kinds of environmental knowledge possessed by indigenous cultures. For those who recognize the importance of these cultures, sustainable development is not so much an invention of the future as a rediscovery of the past, even when the practical mechanics of how to combine modern technologies with a concern for sustainable-livelihood creation remains largely unexplored.

In this chapter the discussion assumes theoretical importance. The development process not only poses problems for sustainability under capitalism, it also poses problems for Marxism, and for intellectual traditions which recognize many of the destructive aspects of global capitalist relations. Marxism begins by asserting a fundamental unity between nature and society, a unity which helps explain the role of labour in transforming the environment, and which also explains how human societies are socially constructed. Central to the Marxist position is the recognition that, as we have seen, capitalism forces commodities to be 'naturalized', to assume a natural quality, while human relations become 'commoditized'. Different forms of social labour transform the environment, and human consciousness is transformed in the process. The difficulty with this theoretical approach is that the superstructural apparatus of developed societies, the framework of law and politics within which the natural environment is exploited, is apparently powerless to prevent threats to nature and the environment which are ultimately threats to production itself. Sustainability is no longer a valuable moral precept alone: it is primarily an essential ingredient in human survival. While the contradictions of capitalism remain unlikely to lead to the immediate destruction of the global economic system, the tensions within this system are increasingly passed on to the poor and powerless and take immediate effect. The ecological crisis, in effect, has overtaken the political crisis in this respect. The widespread poverty and malnutrition that marks sub-Saharan Africa, for example, extends beyond the 'immiserization thesis' in Marx which predicted a *relatively* aggravated position from which class consciousness was capable of developing. In less extreme circumstances than that of Africa the argument still stands: the process which places in jeopardy the livelihoods of the poor does not necessarily lead to a heightened level of consciousness likely to provide solutions to the problem. The attempted viability of underdeveloped economies within the global system is bought precisely at the price of destroying the sustainability of cultures and cultural knowledge. 'Rational' environmental management makes the world safe for development; however, it does not make the environment safe for the poor and their livelihoods.

In looking ahead to the future, then, we need to be conscious of the social and economic consequences of the way nature is being transformed and produced. Later in the chapter the discussion moves on to consider whether the material 'production of Nature', via biotechnology and genetic engineering, provides a way out of the impasse created by the development process. Is the 'production of Nature' in the theoretical debate paralleled by the *actual* production of nature in laboratories and gene banks? Does the potential exist to override many of the problems which this book has debated and exposed through technological means? It is suggested later in this chapter that the very growth of biotechnology and the biological production of nature is a response to the ecological 'crisis' in what Marx termed 'external Nature'. The conclusion then is that research developments which aim to reproduce nature, whatever their productive potential, are unlikely to strengthen the political resolve to find solutions to ecological (and social) problems. On the contrary, the search for 'laboratory' solutions to the problems of conserving the species and increasing agricultural production may well have the effect of hastening the process of environmental degradation, rather than improving it. Just as environmental management needs to be understood as a response to the contradictions imposed by new technologies on the natural resource base, so biotechnology and genetic engineering are technological responses to the limitations of effective environmental management.

The concept of nature in Marx

Marxist writing on the environment owes an enormous debt to the work of Alfred Schmidt, whose book *The Concept of Nature in Marx* (1971) affirms the unity of society and nature in Marxist thought. Written between 1957 and 1960, under the influence of Horkheimer and Adorno, Schmidt's work, like that of other members of the Frankfurt School, seeks to elaborate concepts which lie underdeveloped in the Marxist canon. In his preface Schmidt writes that his intention is to demonstrate that it is the concrete, not the abstract, form of human work which needs to be explained. The 'natural' limits of the dialectic process cannot exist outside material production, outside the labour process (Schmidt 1971, 11). It is necessary to reassert the primacy of labour in the way nature is transformed (1971, 16):

> The commodity as the embodiment of abstract human labour, expressed in units of socially necessary labour-time, is independent of any determination by nature . . . Marx considered nature to be the 'primary source of all instruments and objects of labour' (Critique of the Gotha Programme) that is, he saw nature from the beginning in relation to human activity . . .

In Schmidt's view Marx recognized human dependence on nature. He also fully recognized the priority of 'external Nature' over materially transformed nature (p. 33). Following Hegel, Marx believed that nature stood outside human society in the form of 'first Nature', the nature which had given birth to mankind itself. This point was of great epistemological importance and needed to be reaffirmed. Marx differed from Hegel, however, in the role he ascribed to human labour in this process, and the implications of the transformation of nature for human consciousness.

Marx's approach, as interpreted by Schmidt, lays great emphasis on the social character of nature in human society. In contrast with Feuerbach, Marx regarded nature as anything but passive – nature itself was a dynamic force of enormous potential. Nevertheless, only labour could release this potential. For Marx, nature was mediated by human labour, but at the same time human beings formed part of nature in a holistic sense. This is referred to as 'second Nature' – that is, nature both as an element of human practice *and* the totality of everything that exists. In an historical or chronological sense 'external nature' (first nature) assumed priority in that it was literally 'prior' to man. However, in a theoretical sense 'second Nature' was more important because 'the concept of a law of nature is unthinkable without men's endeavours to master nature' (Schmidt 1971, 70).

The thrust of Schmidt's argument, then, is that since contact with the external world changes human nature 'the dialectic of Subject and Object is for Marx a dialectic of the constituent elements of Nature' (p. 16). His book was written to demonstrate that 'Marxist theory itself already contains the dialectic of nature with which Engels believed it had to be supplemented' (p. 61). In his view human beings change their own nature as they progressively deprive external nature of its strangeness and 'externality', as they mediate nature through their own actions. There is then no contradiction between the interests of nature and human beings, since humans are at once part of nature which is itself drawn into commodity production through the application of human labour.

Schmidt's references to Engels are important, since he was trying to correct what he saw as Engels' distortion of Marx's original purposes in his celebrated essay on *The Dialectics of Nature* written in 1875 and 1876. Engels had identified nature in terms that appeared to separate it off from human beings and asserted that natural forces outside human control also functioned according to a dialectic process. Engels was apparently suggesting 'two different areas of application' for the dialectic, one in 'second Nature' (that is, humanly transformed nature as part of total existence) and the other in first (or external) nature. In Schmidt's view Engels' insistence that there existed natural laws of the dialectic must lead to incompatibility between the dialectic and materialism. He wrote that: 'There can be no question of a dialectic of external nature, independent of men, because all the essential moments of a dialectic would in that case be absent' (Schmidt 1971, 59).

In an extended appendix to his book Schmidt supports Sartre and Hyppolite in arguing that the dialectic should be viewed as a form of motion of *human* historical practice, not as a part of nature 'in itself'. Schmidt was also fully aware of the use that had been made of Engels' position by Soviet apologists, for whom the laws of natural science could easily be subsumed to the laws of dialectic materialism, as during the celebrated controversy surrounding Lysenko and heredity.

Before proceeding to criticize Schmidt's position, it is worth pausing to consider the strength of this very orthodox assertion of Marx's materialism. The unity of society and nature is an important concept to grasp. As Sohn-Rethel (1986) has argued, scientific thought is part of a longer historical process through which human beings acquire social existence in nature. It enables us 'to obtain an idea of what science amounts to for giving mankind the right or wrong productive forces in its existential relation to Nature' (Sohn-Rethel 1986, 115). The quality of our science and of our thinking is a product of material experience. As we saw in the last chapter, Sohn-Rethel distinguishes between 'societies of production' and 'societies of appropriation' in terms of the kind of scientific knowledge which is generated, and the use to which that knowledge is placed in the accumulation process. By abstracting himself from the external world of nature, man the subject came to believe that his new categories of thought were independent of social and historical conditions. This socially necessary consciousness, which is also a 'false consciousness' in Marxist terms, has provided both the epistemological foundations of modern science 'and, on the other hand, has prevented philosophers from recognising the limitations which are inherent in their "autonomy of reason", in virtue of its origins as the ideological reflex of commodity production' (Thomson 1978, 341–2).

Several recent contributions to this debate have emphasized the strength of Marx's analysis, as interpreted by Schmidt. Smith and O'Keefe (1980, 80), for example, assert that 'the unity (but not the identity) of Nature with history underlies all of Marx's work, which ought for that reason alone, to be of critical interest to social scientists'. Burgess (1978) echoes Schmidt's view that the polarization between the naturalization of man and the humanization of nature is a false one. Human needs are not as 'fixed' as those of animals; therefore human exploitation of nature takes a more variable form. As Burgess sees it 'changes in the way man interacts with Nature (that is, changes in the mode of production) are rationalised and expressed as changes in the concepts of nature' (Burgess 1978, 70). In his view, since destructive forces are also productive forces, a perspective that denies the essentially social process through which nature is transformed, is incapable of explaining it.

The primary purpose of Schmidt's work, to assert the unitary character of the 'natural' and the 'social', remains of great relevance to discussions of the environment and development. Problems arise, however, once we begin to analyse the relationship between the environment and the accumulation process itself, to grapple with the business of development and its effects. Schmidt remarks at one point (1971, 17) that 'there is no systematic Marxist theory of nature of such a kind as to be conscious of its own speculative implications', and even if we confine ourselves to 'speculative implications' we soon find Schmidt's own interpretation equally unsatisfactory.

In his examination of Marx's thought Schmidt lays great emphasis on the distinction between use values and exchange values. Use values are taken from Nature but assume 'value' only in meeting human needs. If they are not put to the service of human purposes, use values revert to the natural sphere from whence they came (Schmidt 1971, 73). In effect, 'second Nature' becomes 'first Nature' again.

This approach seriously underestimates the importance of 'external Nature'. Smith and O'Keefe (1980, 83) are right to assert that:

The unreconciled dualism in Schmidt's concept of Nature, particularly the idea of a second and a first nature, are more truly and practically representative of (an) historical period – a period of developing exchange economies – than of any other . . .

This is an important point. Schmidt's distinction, drawing on Marx, fits best when it is applied to highly dynamic periods of commodity production, in which the social identities forged from the production process itself are of primary importance. Table 8.1, which draws liberally on Taussig (1980), illustrates the way in which, during a period of capitalist expansion, the conversion of nature to exchange values materially affects the construction that is placed on nature. Taussig in his work was referring to the point at which pre-capitalist peasant economies in Colombia meet, and apparently embrace, capitalist market relations. As we saw in chapters 6 and 7 the transformation of the environment raises problems for the sustainability of cultures, as well as environments, under these conditions.

Table 8.1 Sustainable production and capital accumulation

	Aim of circulation	*Environmental effects*
Use-value paradigm	to satisfy wants means of exchange	productive capacity
		destructive
Exchange-value paradigm	to gain money as an end/means to make more money	destructive
		productive capacity

In the use-value paradigm the aim of circulation is to satisfy wants and the effect of the irreversible destruction of nature is to destroy value.

In the exchange-value paradigm money becomes a means to acquire capital, and nature yields greater production as it is destroyed. The destruction of nature in unsustainable ways becomes 'natural', productive activity and helps to create value.

Source: adapted from Taussig 1980

There are also other weaknesses in the distinction between use values and exchange values, which are evident the moment we look beyond periods of highly dynamic commodity production. To assert that all nature is altered by human activity does not enable us to distinguish between the different ways in which nature is transformed. In the peasant community discussed by Taussig in Colombia the increasing monetization of the economy, including dependence on wage-labour, served to undermine cultural categories, but the impact of capitalism was not confined to exchange values alone. Similarly, in industrial societies which place a premium on non-productive activity and access to nature, in the provision of public amenity for example, it is difficult to explain why the production of exchange values should be of critical importance. Is the production of nature, as guaranteed by national parks, an exchange value or a use value? Is the commoditization of nature in peasant communities confined to market relations, or does it permeate the household, women's labour and the spheres of domestic reproduction as well as production?

If we are interested, as we should be, in the ways in which the environment is transformed under capitalism, then it might be useful to distinguish between the transformations that occur as nature *enters* the production process (for example, in the tropical forests) and the transformations that occur when natures *leaves* the production process (through soil exhaustion in the Sahel, for example). In this way we might be in a better position to specify the difference that commodity production makes.

The second objection is an even more fundamental one. The distinction between 'first' and 'second' natures is useful in handling Marx's categories, but it does not stand up to close examination. As Smith (1984, 36) argues, 'the production of consciousness is an integral part of (the) general production of material life' but it is a feature of our consciousness of the environment that the value of 'external' (first) nature can also be recognized and acknowledged as influencing productive use. Environmental consciousness is not simply the product of our material involvement in the production of nature, since concepts like that of 'wilderness' play an increasingly powerful role in what we seek to conserve. External nature, such as wildlands, is every bit as much a social construction as 'second Nature', transformed in the process of consumption rather than production. Indeed, society seeks to exercise social control over our access to 'external Nature' and to ensure, in developed countries, that, as we are liberated by technology from subservience to nature, we do not become totally subservient to production. This is the sociological significance of 'conservation'. The problem with emphasizing the unitary character of nature and society in Marx's thought is that 'in Marx nature only appears through the forms of social labour' (Schmidt 1971, 58).

In overcoming the ontological problems of a divorce between nature and society, Marxist thought raises much greater problems. What happens to the environment if the process of transformation cannot be confined to commodity *production*, since the society's very ability to reproduce itself and consume nature is at stake? The reproduction of human society together with the reproduction of nature are both processes which cannot easily be reduced to a division between use values and exchange values. It is precisely at

Table 8.2 Household economy: women's activities and the environment

(1) Market exchanges	(2) Non-market exchanges	(3) Domestic production of use values	(4) Reproduction
(a) Cash from wage labour	(a) Kinship and community networks, payments in kind for services	(a) Agricultural goods, clothing, shelter	(a) Biological reproduction of labour force
(b) Goods for cash	(b) Exchanges of labour or goods		(b) Social reproduction – child care, meeting family needs
(b) Domestically manufactured commodities (clothes, baskets, etc.)			

this point that the theoretical importance of sustainability is evident. The problem with conceptualizing the relationship between nature and society is that both concepts are broader than Marxist theory allows. Nature involves ecological systems which, whatever their relationship to human activity, pose threats to the survival of human society: for example the 'greenhouse effect' in the tropical forests, or the 'nuclear winter' that seems likely to follow a nuclear attack. Similarly, as feminist research has documented, production under capitalism, whether in developed or developing countries, needs to be considered together with biological and social reproduction at the level of family and household (see table 8.2). Since it does not consider these issues, the debate about the mediation of nature also fails to incorporate adequately processes which themselves effect changes in consciousness. It is important to emphasize the role of commodity production as a sphere of production, but the areas that lie outside commodity production are also intimately linked to environmental change and cannot be dismissed so easily. Nature as a constraint on human action and women as the means through which human society reproduces itself cannot be treated as if they were 'outside' the mediation of nature. As Sayer (1983) argues, we can agree that biological powers, for example, are always mediated by social forms, without wishing to argue that they are *reducible* to social forms alone. The environment is more than the production of Nature. The capacity of a system to reproduce itself, biologically as well as socially, rests with the sustainability of its components, at base the family and the resources it commands. These are elements in the production of nature and the reproduction of society which we cannot leave to one side in our analysis.

The industrial appropriation of nature

Until the last hundred years or so the principal constraints on agriculture were environmental: soils, topography, water supply and climate. Crops were produced and animals reared in conformity with the constraints imposed by nature. With the development of modern industrialized agriculture technology has been developed which reduces these environmental parameters, making it possible to alter the conditions under which agriculture is practised by effectively controlling the environment. Mechanical and chemical technologies, as well as developments in food-processing and storage, have increased specialization within agriculture and reduced the importance of location-specific factors in agricultural systems. More recently, genetic technologies have enabled animals and plants to be selected for production, not simply through carefully controlled selective breeding programmes, but also through genetic engineering, altering the very genetic materials that serve to define natural species.

The impact of these new technologies can be reduced to two processes: a progressive appropriation of nature by industrial capital, and the substitution of industrially produced ingredients in food production for those found in nature (Goodman, Sorj and Wilkinson 1987). As Goodman and colleagues see it, capital originally confronted a 'natural production process', in which no industrial alternative existed to the biological transformation of solar energy into food. Agricultural systems had three main components: the 'biological conversion of energy in nature, the growth of plants and animals in 'biological time' (seasons and maturation) and land-based rural activities in space. In the course of little over a century these essential elements in the agricultural system have become progressively appropriated by industrial processes. This process of appropriation is incomplete, since in many cases organic nature and space continue to represent obstacles to the penetration of capital and technology. It has proved impossible to devise a unified production process which replaces nature. However, industrial capitals have responded to the challenge of overcoming environmental constraints by adapting to the specificities of agricultural systems. As Goodman, Sorj and Wilkinson (1987, 1–2) put it:

within the changing limits defined by technical progress *discrete* elements of the production process have been taken over by industry . . . in effect, the process of natural reproduction of plants and animals is being internalised via *science in the reproduction of industrial capitals* . . .

The development of agriculture until the nineteenth century was dependent to a large degree on the mechanization of processes that had formerly been undertaken by hand. This did not remove agriculture's dependence on nature. By contrast handicraft production (not linked so closely to organic nature and space) was transformed to machine production (manufacture) in a much more radical fashion. The mechanization of agriculture, together with some selective breeding of animals, merely served to emphasize the limits beyond which nature could not be transformed.

The growth of plant genetics and crop hybrids, which contributed so importantly to the Green Revolution, was, in its early phase, linked to the chemical revolution in agriculture. Nevertheless these technological developments marked the first real appropriation of natural production processes, serving to reconstitute nature as it was being transformed. Today the processes of industrial appropriation and substitution are more advanced than even 30 years ago, when the research was being undertaken into Green Revolution crops. Today we can also observe a third, related process which is closely associated with industrial appropriation and substitution. This is the process of environmental depletion, under which ecological systems yield genetic materials for gene banks and laboratories, while the systems themselves are broken down and destroyed. Nature is reconstituted not only in laboratory research, through recombinant DNA or tissue culture, but in the genetic collections which are taken from nature and transferred to the laboratory. As the cycles of natural production are modified or disappear in agriculture, so the biological ingredients of ecological systems are conserved outside the land areas and spatial locations where they once existed. Before considering what industry is doing to reassemble nature, through biotechnology and genetic engineering, therefore, we need to examine the process through which genetic materials are being separated off from nature, in gene banks and laboratories.

Banking the species

Writing almost 15 years ago Timothy (1973, 655) stated in unequivocal terms the case for preserving germ–plasm outside its natural context:

> A collection of germ plasm is extremely valuable stuff. The myriad differences are each due to a particular DNA sequence . . . A germ plasm collection preserves the unimaginable combinations of these sequences, sequences which may never be arrived at again, at least until we control DNA synthesis in its entirety . . . Comprehensive collections of germ plasm must be preserved – before the material is lost forever . . .

The struggle to preserve natural germplasm assumes more urgency with the passage of time for two reasons: first, because we are destroying genetic material as we destroy the tropical moist forests which contain an estimated 40 per cent of all the world's species; second, because the selective breeding of crops and animals is so advanced that it seriously prejudices the variety of the species that exist in nature. For a decade now the collection and documentation of crop germplasm has been co-ordinated worldwide by one of the 13 international agricultural research centres, the International Board for Plant Genetic Resources (IBPGR). This organization has overseen field collection of many crop varieties and the establishment of gene banks, where seeds and cuttings from the world's most important food and commodity crops can be stored (Wolf 1985, 238). The largest gene banks are in the most developed countries, the United States, Western Europe, Japan and the Soviet Union, but germ–plasm is also stored in the South on a smaller scale. As Wilkes (1983) makes clear, the establishment of gene banks is a mixed blessing since the fact that 'the centres of genetic variability are moving from natural systems and primitive agriculture to gene banks and breeders' working collections' also implies a concentration of power by concentrating natural resources.

Already political disputes have broken out over germplasm, and there are likely to be many more in the future. In November 1983 the Food and Agriculture Organization (FAO) of the United Nations sought an international undertaking on plant genetic resources at its twenty-second biennial conference. The agreement, which was not binding on the member states, proposed the establishment of a network of local, regional and international centres for collection of germ-plasm. It also suggested putting these collections under FAO auspices (E. C. Wolf 1985). This suggestion immediately met with some opposition, especially from developing countries who perceived in the proposal a covert attempt to control their own genetic resources. Similarly commercial plant breeders' organizations were unwilling to free their highly valued stocks from patents, giving them the same status as the wild species and traditionally cultivated varieties that exist in nature. At issue was not only the control which international organizations were able to exert over developing countries, but also the proprietory rights of transnational organizations in developing countries. The opposition to FAO's proposal was led by a number of countries, including Colombia, Cuba, Libya and

Mexico, which argued that transnational corporations based in developed countries already had virtually free access to their environments, and that the genetic resources they removed were developed into commercial varieties which were sold back to developing countries at considerable profit.

A useful example of what the developing countries fear has recently been provided by MacFadyen (1985) in an illuminating article. He describes what happened when a Florida citrus grower discovered one morning that some of his orange and grapefruit seedlings had developed a new disease. The young trees were afflicted with yellow-green lesions which, laboratory tests revealed, had been attacked by citrus canker in a particularly virulent form. The presence of this canker, which can be transmitted by water, wind or human contact, immediately put in jeopardy Florida's entire $2.5 billion citrus industry. Throughout the Florida citrus belt whole groves of citrus trees were burned to the ground and, within weeks, a major effort was being launched to destroy all suspected citrus plants, numbering over seven million trees.

The outbreak brought draconian measures on the growers because modern agriculture is extremely vulnerable to diseases which threaten to wipe out whole industries. Nobody knew where the canker bacteria originated, or how it found its way to one small grower in Florida, but the reason for alarm can be appreciated when it is realized that 86 per cent of Florida's commercial orange harvest consists of just three genetic varieties. A single strain, Marsh, 'accounted for nearly two-thirds of the state's grapefruit crop' (MacFadyen 1985, 36). Citrus breeders who continually select for ever more desirable commercial varieties have reduced, by their activities, the genetic stock which underpins their industry. Genetic uniformity is the logical consequence of successful production agriculture which requires a very narrow band of varieties for which commercial demand is known to exist. When most farmers were operating under specific environmental conditions disease and pests could be localized, but as agriculture has been developed which can minimize the drawbacks of specific environmental conditions, then the number of producers exposed to new diseases increases. In south China today innumerable citrus varieties grow wild, some of which have probably developed resistance to the canker bacteria which threatened to destroy the Florida crop. By seeking to control the reproduction of these wild varieties the companies and governments involved are able to exercise enormous political power, once they have gained access to natural genetic material. As MacFadyen puts it (p. 38) 'the debate over plant genetic resources is no longer the sole province of technocrats and scientists. Like so many UN struggles, this one pitted the underdeveloped and developing countries of the Southern tier against the developed nations of the North.'

Biotechnology and genetic engineering: 'the genie is out of the bottle'

The successes of the Green Revolution in the 1960s and 1970s, in apparently overcoming environmental obstacles to increased agri-cultural production, were not without cost to the environment. The combination of improved seed varieties, chemical-based fertilizers and pesticides and irrigation, which were the foundations of the successful increases in crop yields, also meant that the new agricultural technologies were not appropriate to the growing conditions encountered by many small farmers, especially those on marginal, rain-fed land. By influencing the farmers' ability to specialize in growing fewer types of crop, under controlled conditions, the Green Revolution also affected the types of food consumed and standards of nutrition. The dramatic changes in land tenure, closely associated with the high-yielding varieties of rice in Asia, were also largely unanticipated by the architects of the Green Revolution in the 1960s.

What we euphemistically refer to as the 'Green Revolution' involved much greater use of chemical-energy inputs, which had the desired effect on yields only when they were applied to the new genetic varieties of wheat and rice. The success of these chemical technologies, then, was intimately bound up with new advances in plant breeding: it was a 'biorevolution' as well as a chemically based 'revolution'. Since 1955, about one-third of the gain in agricultural productivity of the developed countries has also been linked to this process (Brundtland 1985b, 24). The industrialized countries, as we have seen, have relatively easy access to germplasm originating in the South. The apparent success of agriculture in these developed countries owes a considerable amount to the earlier investment in food staples, particularly wheat, in the developing countries. One aspect of this situation, which has already been discussed in chapter

5, is that the disposal of food surpluses in the North (itself partly the consequence of animal- and plant-breeding programmes in the South) now represents a formidable obstacle to sustainable agricultural development in the South.

Apart from the vexed question of germplasm banking, the combination of chemical and biotechnologies has other important implications for agricultural development and the environment in the South. Most chemical-energy inputs are more divisible and less 'lumpy' than other technologies, but control systems are operated from outside the farm and sold to the farmer in the form of packages. The interest of capital lies not only in selling the package to the farmer, however, but also in transforming the product. It is in the transformation of the product that 'value added' is so important within the food system. The process of technological change in agriculture provides for a bigger role for capital in the production process and a reduced one for labour (and independent entrepreneurship). The role of labour, even in so-called family farming is increasingly dictated by technological development within the industrial sector (Goodman and Redclift 1985).

Considering the role of plant and animal genetics in the earlier 'chemical phase', usually referred to as the Green Revolution, it is hardly surprising that agrochemical companies should have devoted so much research time and expenditure to the development of biotechnology. Recent research by Buttel, Kenney and Kloppenburg (1983) points to a growth of concentration within the agrochemical industry, the merger of agrochemical companies and the acquisition of seed companies (1983, 47). The seed industry has become especially important in relation to bio-engineering, and the existence of large, highly diversified 'deep pocket' transnational corporations in the seed industry has proved a stimulant to the successful development of biotechnology research. Oil, chemical and pharmaceutical industries need to expand into seeds if they are to translate the potential of biotechnology into commercial practice. These industries are interested, for example, in developing commercial practices under which transnational corporations in the agrochemical field sell the farmer seeds which have their own built-in, bio-engineered performance. The only alternative to this expansion into biotechnology, for the large agrochemical companies, is to lose the initiative to other transnationals and state-owned companies which are buying up and developing the research knowledge

Table 8.3 Corporate control of the seed market in the United States

Crop	Corporate ownership	% of the market
Beans	Sandoz, Union Carbide, Upjohn, Purex	79
Cotton	Cocker, Pioneer, Southwide, Anderson Clayton	44
Lettuce	Union Carbide, FMC, ITT, Upjohn, Purex	66
Peas	Sandoz, Upjohn	43
Soya	Sandoz, Upjohn, Purex, Shell/Olin, Pfizer, Kent, docker, Pioneer	42
Wheat	Cocker, Ciba-Geigy, DeKalb, Sandoz, Cargill, FMC, Shell/Olin, Pioneer	–
Maize	Pioneer, DeKalb	50

Of the 562 certificates issued by the office before 1979, 46 per cent were to 17 private corporations.

Source: Plant Variety Protection Office (1979), cited by Mooney 1979.

acquired by small, venture capital organizations. Table 8.3 shows the extent to which the seed market in the United States is already effectively under corporate control.

The impetus to the development of biotechnology also comes from another quarter: the chemically based technologies are heavily reliant on petroleum, and after the rise in oil prices during the early 1970s it was attractive to seek new sources of energy which used the biomass, rather than fossil deposits, as the energy source. It became recognized that in future food development the costs of energy would assume greater importance. The question was asked: 'how many calories do we get back for each one that we put in?' (Apple 1982, 478). As we saw in chapter 2, relatively 'primitive' agricultural systems are frequently more efficient than industrialized agriculture in their use of energy. In the developed countries, writes Apple, 'we get back between two and three calories for each one that we put in, and in some instances, such as for tomatoes in California, we get back 0.6. We get back less than we put in' (p. 478). Genetic engineering, in plants and animals, provides the possibility of rising to the challenge of the global food crisis. The more resources devoted to genetic technology, it is argued, and the sooner its benefits are made available, the sooner the food problems of developing countries will be conquered.

During the 1970s it also became apparent that the high-yielding varieties of wheat and rice had brought in their train a new series of environmental problems, on a scale that had not been appreciated. In many respects the industrialization of agriculture had brought problems of scale and pollution which were not usually associated with farming activities. As agro-industrial enterprises became larger it became easier to pass on pollution problems to others, and increasingly difficult to define the limits of the corporations' responsibility. Large agricultural projects required extensive irrigation and the control of crop pests by the use of chemical pesticides. The project, once started, promised to bring employment to the area and the population in the vicinity of the project increased. However, large irrigation schemes also increased the risk of water-borne parasites, and the use of pest-control techniques made insect-disease vectors resistant to insecticides. Some authors conclude that the net result could well be a worsening of the quality of life, rather than an overall improvement (Bishop and Cook 1981). The real environmental cost of development is apparent only some time after the project is initiated. Throughout the 1970s, and in the face of impressive production increases, it became clear that the environmental costs of 'advanced' agriculture, based on chemical-energy sources, had increased. The attraction of biotechnologies for policy makers and corporations was that energy was produced in a renewable, generally non-polluting way, from 'natural' resource stocks, while the control of technology could still be maintained by the scientists, or the company employing them.

The appeal of biotechnology research and genetic engineering was that they promised fewer environmental 'externality' effects than conventional petroleum-based agricultural technology, at what would ultimately be a lower cost. They also overcame resource scarcities, especially the perceived scarcity of arable land. Industrial biotechnology could replace the land base by partially removing production from land. However, it could *also* enhance the prospects of certain agricultural commodities by providing new uses for them, for example, starch and sucro-based products. The Green Revolution had worked best on good land and seemed to offer little to farmers on poorer or marginal land. The thesis grew that, since there was not enough good land, and food shortages were still a pressing concern in much of the South, relatively poor land could be turned over to the new biotechnologies for the production of a feedstock. Martin Apple (1982, 478) wrote:

> We are on the threshold of the transformation of plants into novel varieties with important new properties. By adding new genetic information to plants and by altering the genetic information they already contain, the possibility of creating food plants which can overcome the barriers we now see as constraining our food supplies is evident . . .

A comparison of Green Revolution technology with that of the new biotechnologies is presented in table 8.4, taken from Kenney and Buttel (1985). This table brings together information about the institutional aspects of biotechnology and its point of insertion within society which are frequently ignored in the voluminous technical literature.

Apart from the problems with chemical-led agriculture, and the scarcity of resources which could be allocated to future versions of the Green Revolution, biotechnology had a number of technical features and possibilities which augured well for its development. The development of molecular genetics in the 1950s and early 1960s had served to link food science with pharmacology and plant genetics to an unprecedented degree. There was common ground, as never before, in the applications of scientific knowledge. In particular, plant genetic engineering had advanced, or was about to advance, in three important directions. The use of recombinant DNA enabled the genes to be 'split'. Philip Leder, in a recent paper entitled 'Moving genes: promises kept and pending', has argued that recombinant DNA engineering had already fulfilled its expectations in the field of mammalian genetics, the promise that was *pending* was the extension of this research to a wider variety of molecular subjects. He also argued forcefully that genetically engineered varieties of living organisms should be returned to nature 'to understand the nature of control, we must have the ability to manipulate a gene in its natural context' (Leder 1982, 39). To test genetic manipulation properly, he argues, it is necessary to reintroduce new, genetically engineered material into nature.

Other breakthroughs in biotechnology and genetic engineering seem to promise as much as recombinant DNA. One of these is

biological nitrogen fixation. Since 1973 research has been undertaken into ways of substituting chemically fixed nitrogen by a
biological process. This would allow micro-organisms to fix atmospheric nitrogen, reducing it to ammonia, which could then be
assimilated by plants. The benefits of this process would be several,
especially under circumstances where nitrogen fertilizers could not

Table 8.4 A comparison of the institutional structures of the Green
Revolution and Biorevolution

Characteristics	Green Revolution	Biorevolution
Crops affected	Wheat, rice, maize	Potentially all crops includ-ing vegetables, fruits, agro-export crops (e.g. oil palms, cocoa) and speciality crops (e.g. spices, scents)
Other products	None	Animal products Pharmaceuticals Processed food products Energy
Areas affected	Some less developed countries; some locations (i.e. if accompanied by irrigation, high-quality land, transport avail-ability, etc.)	All areas; all nations; all locations, including marginal lands (character-ized by drought, salinity, Al toxicity, etc.)
Technology development and dissemination	Largely public or quasi-public sector	Largely private sector (multinational corporations and start-up firms, with the former predominating in terms of commercialization)
Proprietary considerations	Patents and plant variety protection generally not relevant	Processes and products patentable and protectable
Capital costs of research	Relatively low	Relatively high
Research skills required	Conventional plant breeding and parallel agricultural sciences	Molecular and cell biology expertise plus conventional plant breeding skills
Crops displaced	None (except the germplasm resources represented in traditional varieties and landraces)	Potentially any

Source: Kenney and Buttel 1985.

be obtained. Similarly, research has addressed the problem of genetic resistance to pests, in the hope of enabling genetic engineering to do what pesticides currently do, without endangering local ecosystems or damaging the environment. Breeding genetic resistance in plants remains one of the major challenges of the new technologies. In future it is likely to become much easier to programme 'against' nature, by using plant breeding to adapt agriculturally important crops to environments which are usually considered unsuitable or hostile to their growth.

The assault on genetic engineering

The most outspoken critics of genetic engineering, such as Jeremy Rifkin, argue that recent breakthroughs in the field amount to a second 'genesis' which we are unlikely to be able to control:

> With recombinant DNA technology it is now possible to snip, insert, stitch, edit, program and produce new combinations of living things just as our ancestors were able to heat, burn, melt and solder together various inert materials' (Rifkin 1985, 41).

Unlike previous developments in genetics, new-style genetic engineering involves manipulation not at the level of the species, but at the genetic level itself. As Rifkin says, 'the working unit is no longer the organism, but rather the gene' (p. 42). In his view, when we have created nature, we soon lose respect for it, we lose our feeling for nature and our place within it.

This kind of doubt about the ethical value of genetic engineering is allied to a number of more concrete objections, which Rifkin and others have made. The principal objections are:

1 That genetic engineering, by increasing the rapidity of photo-synthesis, places an unacceptable burden on an already overtaxed soil structure.
2 That genetically engineered species, if reintroduced into the environment, pose problems for the ecological systems to which they have not adapted in the usual 'evolutionary' way.
3 That genetic diversity will decline as only the most commercially attractive plants and animals are reproduced.
4 That the control of the species in gene banks becomes easier when

the genetic material can be engineered. There is thus a prospect of tighter and more monopolistic control of nature.

5 That it is possible, and for some attractive, to use recombinant DNA techniques for warfare. According to Rifkin, 'scientists say that they may be able to clone selective toxins to eliminate specific racial or ethnic groups whose genotypical make-up predisposes them to certain disease patterns' (Rifkin 1985, 58). 'Enemy' economies could thus be destroyed, theoretically, by destroying their food systems.

6 Finally, that the patenting of biotechnology and genetic engineering is against the interests of the public, since freedom of information is denied them.

Not all critics would agree with each of these objections to genetic engineering, and most would place more emphasis on one or other of the objections listed above. Nevertheless, most critics link a number of these points, and even the most moderate of them have serious reservations about one or other field of research. At the same time, the most persuasive defenders of biotechnology and genetic engineering, such as Elkington (1985) became interested in the subject precisely because of their own doubts about the risks to human health and the environment of industrial and chemical technologies (1985, 12). In his latest book (1986), Elkington documents some of the evidence that human reproductive systems are under attack from pesticides, industrial wastes and other chemicals. Objections to the harmful environmental impacts of non-biological technologies have served to make him a firm advocate of genetic engineering, as less socially harmful and environmentally damaging than most of the technologies which preceded it.

Some scientists draw distinctions between the different depths at which genetic engineering is developed. Starlinger (1982), for example, distinguishes three main areas in which genetics has been brought to bear on biotechnology: the production of natural compounds using genes cloned in bacteria or yeasts; the introduction of genes into somatic cells and the introduction of genes into the germ line. He draws the ethical line after the second of these (1982, 233):

It is very important to make an absolutely clear distinction between genes being introduced into somatic cells and into the germ line. If genes are introduced into somatic cells, this has no influence on future generations. It is only when we go to the germ line that such long-lasting effects will be exerted.

Similarly, some scientists believe that the introduction of new genetically engineered varieties into nature is not necessarily injurious to the environment. Nossal (1985, 126) responds to the criticisms launched by Rifkin and others by commenting that:

> even the worst examples of what evolution can do to place the ecosystem at the mercy of a single species reach their limits in space and time. The plague did not destroy Western civilization. The grasshopper did not put an end to planned agriculture . . . I cast doubt upon the capacity of any (genetically engineered) species to wreck the world. The feedback loops within the ecosystem are too many and varied for that to happen.

One of the most insistent criticisms of the unchecked expansion of biotechnology and genetic engineering is that taking out proprietary rights in nature seriously jeopardizes freedom of information and knowledge. On 16 June 1980 the United States Supreme Court ruled by a majority of five to three that General Electric had the right to patent a 'superbug' which, when contained in straw, could be inoculated into oil spills, enabling the spills to be literally eaten by a bacterium called *pseudomonas*. Speaking for the majority Chief Justice Warren argued that 'the fact that micro-organisms, as distinguished from chemical compounds, are alive is a distinction without large significance'. Thenceforth biotechnology was handed over to the private corporations and their legal representatives. This prompted Cherfas (1982, 204) to comment that:

> Simply allowing academics to get on with it and invest their expertise in industry solves very few problems. In particular, it does nothing about the conflicts of interest that must inevitably exist.

Jonathan King, an MIT molecular biologist who has opposed unregulated recombinant DNA research, argues that the interests of the corporation and those of the scientific community are opposed. The scientific community needs free access to information to make progress, while the corporation needs, at least initially, to protect its products by secrecy and patenting. King reported at a meeting of the American Association for the Advancement of Science in Washington, that a vice-president of the Exxon Corporation had advised Exxon scientists attending a professional conference that their presentations needed to be vetted by the Corporation before they attended, and urged them not to speak to anybody on their way there! King argues that genetic engineering should be like nuclear weapons, in that private corporations should not be allowed to sell the product, even in a very profitable market (Cherfas 1982, 206). Whatever the eventual outcome of the debate about patenting biotechnology and genetic engineering, there is no evidence that at the moment commercial enterprises treat genetically engineered products any differently from other products, such as drugs or chemicals (Nossal 1985). The genie may be out of the bottle, but the laboratory is being subjected to increased corporate control.

Sociobiology: conferring legitimacy on nature

In chapter 1 it was suggested that the historical development of the social sciences indicated not so much a conclusive rejection of organic and evolutionary theory as a continuing tension between two scientific traditions: one of which sought explanations for social structure in nature, internalized by human beings in the course of evolution, while the other tradition referred to the growth of culture and social complexity within human society itself, outside the individual as well as inside. Social theory, from the late nineteenth century onwards, emphasized the social against the natural, organically based tradition. By the middle of this century evolutionist and biologically grounded theories of human society had been effectively discredited, at least in the eyes of most social scientists. Although recourse was made to naturalistic analogies and metaphors, explanatory theory in the social sciences was largely taken up with finding non-naturalistic causes for progress in human societies.

The development of the biological sciences in the last two decades has ensured that non-naturalistic explanations for human society did not disappear; it now appears they were merely held in abeyance during the high tide of sociological explanation. The growth of biotechnology and genetic engineering during the last decade or so has helped to rehabilitate biological determinism in a new guise which is more sophisticated, and less easily dismissed, than earlier theories. Since the publication of E. O. Wilson's *Sociobiology: the New Synthesis* in 1975, the world has had to live with a new representative of the older organic tradition in the form of sociobiology.

In his first book on sociobiology Wilson (1975, 4) defined his approach largely in terms of animal societies:

> the systematic study of the biological basis of all social behaviour. For the present it focusses on animal societies . . . But the discipline is also concerned with social behaviour of early man and the adaptive features of organization in the more primitive human societies . . .

This modest attempt to bring together traditions in ethology and physical anthropology with those of biology did not remain modest for very long, however. Three years later Wilson (1978, 16) was writing:

> What is truly new about sociobiology is the way it has extracted the most important facts about social organisation from their traditional matrix of ethology and psychology and reassembled them on a foundation of ecology and genetics studied at the population level in order to show how social groups adapt to the environment by evolution . . .

Evidently, the study of animals and early hominids could be extended to all human societies, by retaining the evolutionary perspective. Wilson was in the process staking out new ground which could not avoid being heavily contested.

The reference to the 'environment' at the beginning of *On Human Nature* (1978), however, failed to disguise the fact that, in Wilson's view, genetics played at least as important a role in shaping human behaviour as environmental factors. Indeed, genetics in his view 'already is decisive' (1978, 19) in shaping human behaviour, whatever the accumulated evidence about non-hereditary components. Sociobiology provided a supportive structure for developments in eugenics and biological engineering, which had revolutionized the natural sciences and served to reaffirm the centrality of genetics in the sciences.

As Rose, Lewontin and Kamin (1984, 236) show, 'the central assertion of sociobiology is that all aspects of human culture and behaviour, like the behaviour of all animals, are coded in the genes and have been moulded by natural selection'. Sociobiologists prevaricate on how far genetics determines detailed human behaviour, but they insist that ultimate control lies with genetic factors. The appeal of this position is not difficult to grasp. On the one hand, sociobiology, by arguing that present social arrangements are genetically determined, serves to legitimize the status quo. Its claims to be a non-political science only make this legitimation more effective. On the other hand, sociobiology appears to take account of what are unquestionably important scientific breakthroughs in the area of genetics, enabling its defenders to argue that it is a serious attempt to extend to the social sphere of behaviour some of the insights gained from science. Sociobiology is presented as if it were a 'natural' science, partly *because* social science frequently implies difficult ethical and political stances.

There are several facets of explanation in sociobiology which deserve attention:

1 Sociobiology is reductionist (Rose, Lewontin and Kamin 1984, 236). Its adherents claim that the characteristics of human behaviour can be reduced, in large measure, to the action of genes. This genetic component has been subjected to natural selection, helping the strongest to survive the process of evolution. The appeal of the theory is that human society as we know it can be described both as inevitable and as the outcome of an adaptive process in nature.

2 Sociobiology never effectively distances itself from the naturalistic fallacy that what exists is therefore desirable, that 'ought' statements can be derived from 'is' statements. Although Wilson has denied that the naturalistic fallacy lies behind many of the statements of sociobiology, it is clear that the enthusiastic espousal of genetically grounded explanations implies, and sometimes explicitly states, that our own selection (and hence our behaviour) has been for the best.

3 One of the implications of believing that human behaviour is genetically determined is that, if we do wish to change human behaviour, we will be most successful if we seek to change the genes.

For this reason Rifkin (1985) argues that sociobiology, allied to genetic engineering, is likely to advance the cause of selective human breeding, eugenics. There are certainly signs of this in some of Wilson's writing:

> The human species can change its own nature. What will it choose? Will it remain the same, teetering on a gerrybuilt foundation of partly obsolete Ice Age adaptations? Or will it press on towards still higher intelligence and creativity, accompanied by a greater – or lesser – capacity for emotional response? New patterns of sociality could be installed in bits and pieces. It might be possible to imitate genetically the more nearly perfect nuclear family of the white-handed gibbon or the harmonious sisterhoods of the honeybee (Wilson 1978)

Where sociobiology differs from earlier versions of biological determinism is in its emphasis on natural selection through the gene, rather than through whole social groups. The evolutionary theorists of the past emphasized the importance of selection in human society but could not bring modern genetic knowledge to their aid in asserting this view. Moreover, as Rose, Lewontin and Karmin show (1984, 238) sociobiologists use optimality arguments derived from economics, claiming that human beings are inherently problem solvers who choose strategies for the optimal solution to environmental problems.

4 The appeal of sociobiology is therefore that it claims universal characteristics for human behaviour, that these universals are innate and yet have evolved to play the role they play through natural selection. Problems inevitably arise in explaining the enormous range of human behaviour, largely because sociobiologists refuse to confront the problem of ideology and culture. It is assumed that certain characteristics such as aggression, entrepreneurship and dominance are characteristics of all societies, without seeking to locate these characteristics within an historical and cultural milieu. They choose to separate ideological and social constructs from their structural roots, in peoples' minds and in human experience that is learned rather than inherited. Areas of environmental adaptation, which have also enriched human consciousness and knowledge, need to be related to historical accounts of culture. It has been argued in

198 SUSTAINABLE DEVELOPMENT

this book, for example, that we might benefit from the knowledge that shifting cultivators or peasants possess about their environments. Sociobiologists might regard such knowledge as somewhat esoteric and irrelevant, since the societies of shifting cultivators and peasants are faced with imminent collapse. As long as we survive genetically, does it matter that our symbiotic relationship with nature, the hallmark of sustainability, has died? The implication of taking a narrow, reductionist view of progress is that we lose variety in human culture, just as we lose variety in living species.

This chapter has examined the ways in which the frontiers of sustainability are being extended. It has considered the ethical and political problems which accompany the development of biotechnology and genetic engineering. If we begin with the knowledge that human action can imperil the survival of the species, several courses of action are possible. We can either seek a renewable resource base, perhaps through biotechnology, which will assist us in conserving endangered species, or we can develop new genetically engineered possibilities of withstanding environmental and human stress. The fact that sustainable development has new technological frontiers does not mean that it is any easier to live within them. We began by looking at the way in which capitalist development had served to transform the environment, in time and space. We should end by returning to the contradictions of this development process and examine what we can learn from them.

9

Conclusion

Exploring the relationship between the environment and development has proved to be a complex, but rewarding, enterprise. In examining the concept of sustainable development it was suggested that the term could express more than a pious hope, but rather less than a rigorous analytical schema. Sustainable development is a concept which draws on two frequently opposed intellectual traditions: one concerned with the limits which nature presents to human beings, the other with the potential for human material development which is locked up in nature. Unravelling and deconstructing this contradiction has been a principal focus of this book. Sustainable development, if it is not to be devoid of analytical content, means more than seeking a compromise between the natural environment and the pursuit of economic growth. It means a definition of development which recognizes that the limits of sustainability have structural as well as natural origins.

It was suggested that the problem in achieving sustainable development was related to the overriding structures of the international economic system, which arose out of the exploitation of environmental resources, and which frequently operate as constraints on the achievement of long-term sustainable practices. At the same time, one of the reasons we are not better equipped to take sufficient account of ecological aspects of economic growth and development is because the intellectual traditions which we draw upon for solutions to problems point in different directions. Neither neo-classical nor Marxist economics take sufficient account of the environment, while environmentalist positions provide only the vaguest guidelines for negotiating a more constructive relationship with nature. Indeed, the argument could be put that containing economic demands for material advance, in a highly unequal world, requires political measures that are so authoritarian they would immediately contradict the liberating, humane objectives that would make development sustainable in the first place.

The first contradiction that lies dormant within 'sustainable development', then, is one which we ignore at our peril: if we cannot rely upon market forces to sustain our environment, we need to place very much greater reliance on international agreement and planning, without which individual, personal or national, interests will dictate the course of the development process. In discussing environmental management it was argued that most planned, environmental interventions are quite unlike this. Most interventions in the development process on behalf of the environment are motivated by a desire to minimize the 'externality' effects of development, rather than to provide lessons in how development should proceed. Where environmental considerations clash with strategic, political or national interests, they are unceremoniously forgotten. We also saw that development policy and practice, often unwittingly, has environmental effects that are indirect and consequently ignored. The debt crisis in Latin America and Africa today is a vivid illustration of a problem with serious environmental implications and causes which is routinely considered in exclusively economic, even financial terms.

A second contradiction concerns the relationship between the political struggles over the environment in developed and developing countries. In seeking sustainability in the North we are seeking to affirm a cluster of related values, concerning the way in which we want our environment to be preserved. We seek, with millions of other people in the developed world, to protect and conserve rural space, to recognize aesthetic values in the countryside, to provide better access to this space and to ensure the biological survival of threatened species. Environmental objectives in the South are rather different. The survival of the species is equally important, although possibly for more crudely economic reasons. Otherwise the environment is contested for different reasons in developing countries. The environment, especially the rural environment, is a contested domain in the South because it is the sphere in which value is created through the application of human labour to nature. If people are to increase their share of material rewards in developing countries, it follows that they must extend their control over the environment, or over the way in which technology transforms the environment.

At this point it is important to remember that the environment has an international character. As Trainer (1986) has observed, the material standards of life in the developed countries are intimately linked with the way resources and human labour are exploited in the South. It is not left to people in developing countries to decide how to use their resources. A reduction in the amount of work, for example, in the industrialized countries can be hailed as a step on the way to more ecological solutions (Ekins 1986). For over a century socialist and social democrat movements have sought to provide better living standards by reducing the duration and intensity of work. If development in post-industrial society means a reduction of labour effort, it may meet the needs of those working-class people in developed countries who are spared the ravages of full-time unemployment. But will a workless future, based on international specialization in technology and relatively high living standards, meet the needs of millions of poor people in the South, where the other face of development frequently implies reduced control over resources for greater labour effort? As Michael Barratt Brown recently put it (1984, 106), the part of the Green case which is most convincing, in relation to traditional Marxist thinking, is that today:

> Capital accumulation means reducing costs by exploiting labour, exploiting all natural resources, exploiting the biosphere, and in the end *without* the result that Marx hoped for – the Revolution that would establish a socialist commonwealth.

Whether or not development is *necessarily* unsustainable, as Trainer (1986) and Bahro (1982) argue, it is clearly unsustainable on current models for many of those whose livelihoods are made in the South, and for reasons that lie outside their control.

Exploring the contradictions of sustainable development also has implications for several other areas of thinking and practical policy-making. It has implications for three in particular: the view that we hold of 'environmental rationality', the changing role of technology in our relationship with the environment and the body of social theory that can help us interpret and understand this relationship. Let us take each of these areas in turn.

As Michael Thompson has recently argued (in Thompson, James and Taylor 1985; Thompson 1986), different perceptions of the environment are neither more nor less 'rational' – they merely reflect the way in which we look at the world. One person's world of resource depletion is another person's world of resource abundance. Consequently divergent views are not necessarily correct or false and are unlikely to be consistent as long as people have different interests and different sources of knowledge and information.

> What we have is not the real (environmental) risks versus a whole lot of misperceptions of those risks but the clash of *plural rationalities* each using impeccable logic to derive different conclusions (solution definitions) from different premises (problem definitions) (Thompson 1986, 2).

Thompson argues forcefully for a greater recognition from mediating institutions, such as management, that there is no single solution as long as different rationalities exist. In recognizing the pluralities of rationality that govern environmental thinking, he says, we are taking an important step towards better decision-making.

Without disagreeing with Thompson about the diversity of rationalities, and the need to recognize their separate strengths, it needs to be emphasized that the strengths of each position are closely related to the economic and political realities that underscore their acceptance by development and environmental institutions. If environmental management was reconstructed to take more account of indigenous knowledge, as argued in chapter 7, then we might reasonably expect differences in environmental rationality to provide creative tensions, from which better policy could emerge. However, it is also important to acknowledge that environmental rationalities are not only socially constructed, they are also supported by social groups with different degrees of power and with conflicting economic interests. The substantive concerns of people with their environment, I have sought to argue, are closely related to the historical processes through which the environment is transformed. Our first task then must be to recognize the links between environmental rationality and the political economy which has contributed to its formulation and intellectual genesis.

The second important area for which sustainable development has implications is in the way technology mediates our relationship with the environment. As we saw in chapter 8, the frontiers of sustainability are constantly shifting. Developments in biotechnology, for example, leave open the possibility that resources can be produced from nature without permanently harming the biosphere. There is nothing inevitable about the destructive progress of science. As Einstein wrote, 'we should be on our guard not to overestimate science and scientific methods when it is a question of human problems, and we should not assume that experts are the only ones who have a right to express themselves on questions affecting the organization of society' (Einstein 1949, 215).

It was argued in chapter 8 that, if we view biological possibilities for creating nature too instrumentally, we risk losing a sense of the wider time horizons which govern the evolution of the environment. We are offered technological breakthroughs not as a way of resolving the contradictions of development for the environment, but as a way of distancing ourselves from these contradictions. The technical 'fix' of genetic engineering needs to be considered alongside the measures which could be taken to conserve the biosphere, not as an alternative to environmental conservation. Technological solutions have a large role to play in the conservation of the biosphere, but the danger today is that the inability to 'manage' the environment in the South, itself partly a consequence of the development process, will lead policy-makers and governments to a technological option that they *can* control, without counting the environmental or social costs. Perhaps we are in danger of thinking about genetics as the nineteenth-century colonial adventurers thought about the cultures they conquered and risk losing, in the process, both the real co-operation of those who depend upon the environments and their environmental rationality. If research and development were carried out by small farmers, for example, rather than by corporate capital, is there any reason to believe that biotechnological research would not be more environmentally and socially sensitive?

The third main area that we need to address is that of social theory. As we saw in chapter 1, since the nineteenth century biological metaphors have been used in the social sciences, but the relationship between the biological and the social in our behaviour has often been avoided by social science. The ascendancy of social theory was closely associated with discredited biological determinism, only to reappear today in a more sophisticated guise as 'sociobiology'. One of the central contradictions in the discussion of sustainable development today relates to this history, for the failure of political economy to address environmental issues cannot be separated from its intellectual inheritance in the nineteenth century.

At the same time the claims of sociobiology and other similar positivist positions are made more plausible by the failure of social theory to address the wider parameters of economic and social behaviour from within an historically grounded and international perspective. A social theory that is turned in on itself will not explain the central paradox that haunts this book: that through the use of methodologies developed in the natural sciences nature has been divested of social control. We are losing control both of the destruction of nature and its recreation. In exploring the contradictions of sustainable development we are necessarily led to explore this paradox.

This book has discussed numerous aspects of development which present barriers to sustainability. It has been argued that scientific knowledge is developed in ways that make it difficult to assimilate and utilize the experience and epistemology of poor people who depend upon the environment for their survival. At the same time decisions are made about resources and the environment at the margins of government departments, beset with political rivalries and sectoral myopia. These decisions are conveyed to countries without real political weight within an international community whose principal members, the OECD countries and the Eastern Bloc, practise policies that are themselves unsustainable, even in narrow, national terms. The attachment to technological 'solutions' is explicable, then, in terms of the failure to derive relevant knowledge from relevant practice and to exercise the political will necessary to generate global recovery. Anticipatory environmental planning and regulatory practice have a role to play in ameliorating the effects of development contradictions, but they cannot hope to overcome the barriers to sustainability contained in current development practice. Unless we are prepared to interrogate our assumptions about both development and the environment and give political effect to the conclusions we reach, the reality of unsustainable development will remain, and the risk of ecological destruction will increase where it is already most pressing. Sustainable development is founded upon a contradiction, with which this book opened. Just as we are poised to unlearn the lessons of the development process we are faced with the possibility that sustainability itself will be put in jeopardy, by leaving it to scientists alone to explore. In seeking to make development sustainable we might begin with our own assumptions and our own practice.

Bibliography

Apple, M. (1982) in F. Ahmad, J. Schultz, E. Smith and W. Whelan (eds.) *From Gene to Protein: Translation into Biotechnology*, London, Academic Press.

Bahro, R. (1982) *Socialism and Survival*, London, Heretic Books.

Baker, L. (1981) 'The environmental nexus', *Resource Management*, 13(2): 12–25.

Barbira-Scazzochio, F. (1980) (ed.) *Land, People and Planning in Contemporary Amazonia*, University of Cambridge.

Barratt-Brown, M. (1985) *Models in Political Economy*, London, Penguin.

Bartelmus, P. (1986) *Environment and Development*, London, Allen and Unwin.

Batisse, M. (1985) 'Action plan for biosphere reserves', *Environmental Conservation*, 12(1): 17–27.

Bernstein, H. (1979) 'African peasantries: a theoretical framework', *The Journal of Peasant Studies*, 6(4).

Bishop, J. and Cook, L. (1981) 'Genes, phenotype and environment' in J. Bishop and L. Cook (eds.) *Genetic Consequences of Man-Made Change*, London, Academic Press.

Blaikie, P. (1985) *The Political Economy of Soil Erosion in Developing Countries*, London, Longman.

Block, H. R. (1981) *The Planetary Product in 1980*, Washington, US Department of State.

Blowers, A. (1985) 'Environmental politics and policy surrounding minerals, agriculture, air pollution and nuclear waste', paper presented to RESSG Conference 'Environmental Problems and Politics in Rural Societies', Loughborough.

Booth, D. (1984) *Marxism and development sociology: interpreting the impasse*, mimeo.

Bowler, I. (1985) *Agriculture Under the Common Agricultural Policy*, Manchester University Press.

Brandt Commission (1983) *Common Crisis*, London, Pan Books.

Branford, S. and Glock, O. (1985) *The Last Frontier*, London, Zed.

Brown, L. (1984) *The State of the World*, New York, Worldwatch Institute, W. W. Norton.

Brundtland (1985a) *Mandate for Change: Key Issues, Strategy and Workplan*, World Commission on Environment and Development, Geneva.

Brundtland (1985b) Brundtland Commission Public Hearings, Jakarta, transcript.

BTAM (1985) British Tropical Agricultural Mission, Bolivia, Review, London, Overseas Development Administration.

Bull, D. (1982) *A Growing Problem: Pesticides and the Third World Poor*, Oxford, Oxfam.

Burbach, R. and Flynn, P. (1980) *Agribusiness in the Americas*, New York, Monthly Review Press.

Burcham, T. (1957) *California Range Land*, Sacramento, California, Department of Natural Resources, Division of Forestry.

Burgess, R. (1978) 'The concept of nature in geography and Marxism', *Antipode* 10(2): 1–11.

Burton, D. J. (1981) 'The political economy of environmentalism', *Kapitalistate*, Working Papers 9, 147–57.

Buttel, F. (1983) *Sociology and the Environment: The Winding Road toward Human Ecology*, Cornell University, Department of Rural Sociology.

Buttel, F., Kenney, M. and Kloppenburg, J. (1983) *Biotechnology and the Third World: towards a global political–economic perspective*, Cornell University, unpublished MS.

Caufield, C. (1984) *Tropical Moist Forests*, London, Earthscan, IIED.

CEPAL (1985a) *The Environmental Dimension in Development Planning: main issues in Latin America*, Santiago, Chile, CEPAL (United Nations).

CEPAL (1985b) 'El Medio Ambiente como factor del Desarrollo', *Notas sobre la Economia y el Desarrollo*, CEPAL, 417, May.

Chambers, R. (1986) 'Sustainable livelihoods', Institute of Development Studies, University of Sussex, mimeo.

Cherfas, J. (1982) *Man-Made Life*, Oxford, Blackwell.

Clements, F. (1916) *Plant Succession*, Washington, Carnegie Institute.

Colchester, M. (1986) 'Unity and diversity: Indonesian policy towards tribal peoples', *The Ecologist*, 16(2/3), 89–98.

Commoner, B. (1971) *The Closing Circle*, New York, Knopf.

Conlin, S. (1985) 'Anthropological advice in a government context', in R. Grillo and A. Rew (eds.) *Social Anthropology and Development Policy*, London, Tavistock.

Conway, G. (1984) *Rural Resource Conflicts in the UK and Third World – Issues for Research Policy*, London, Imperial College/SPRU, Papers in Science, Technology and Public Policy.

Conway, G. (1985a) 'Agro-ecosystem analysis', *Agricultural Administration* 20, 31–55.

Conway, G. (1985b) 'Agricultural ecology and farming systems research', paper prepared for the Farming Systems Research (FSR) Workshop, Hawkesbury, Australia.

Conway, G. (1985c) 'Rapid rural appraisal and agro-ecosystem analysis: a case study from Northern Pakistan', paper presented at International Conference on RRA, Khon Kaen, Thailand.

Cook, K. (1983) 'Surplus madness', *Journal of Soil and Water Conservation*, 31(1), 25–8.

Cotgrove, S. (1982) *Catastrophe or Cornucopia: The Environment, Politics and the Future*, Chichester, Wiley.

Crow, B. and Thomas, A. (1982) *Third World Atlas*, Milton Keynes, Open University Press.

Dandler, J. and Sage, C. (1985) 'What is happening to Andean potatoes? A view from the grassroots', *Development Dialogue*, 1, Uppsala, Sweden.

Dasmann, R. F. (1975) *The Conservation Alternative*, London, Wiley.

Dasmann, R. F. (1985) 'Achieving the sustainable use of species and ecosystems', *Landscape Planning* 12: 211–19.

Denevan, W., Treacy, J., Alcorn, J., Padoch, C., Denslow, J., Flores, S. (1982) 'Indigenous agroforestry in the Peruvian Amazon: Bora Indian management of swidden fallows', in J. Hemmings (ed.) *Change in the Amazon Basin*, Manchester University Press, vols. 1 and 2.

Devall, B. B. (1979) 'The Deep Ecology Movement', *Natural Resources Journal*, 20, 299–322.

Devall, B. B. and Sessions, G. (1984) *Deep Ecology*, Layton, Utah, Peregrine Smith Books.

Diaz, Bernal (1963) *The Conquest of New Spain*, London, Penguin.

Dunbar Ortiz, R. (1984) *Indians of the Americas*, London, Zed.

Durkheim, E. (1964) *The Division of Labour in Society* (1893), New York, Free Press.

Ehrlich, P. (1974) *The End of Affluence*, New York, Ballantine Books.

Einstein, A. (1949) 'Why Socialism?' in Levidow, L. (ed.) (1986) *Radical Science Essays*, London, Free Association Books.

Ekins, P. (ed.) (1986) *The Living Economy*, Routledge and Kegan Paul.

Elkington, J. (1985) *The Gene Factory – Inside the Biotechnology Business*, London, Century Publishing.

Elkington, J. (1986) *The Captive Womb*, London, Penguin.

Ellison, L. (1954) *Subalpine vegetation of the Wasatch Plateau*, Utah, Ecological Monograph 24, 89–184.

Engels, F. (1970a) 'Introduction to the Dialectics of Nature' in K. Marx and F. Engels, *Selected Works*, London, Lawrence and Wishart.

Engels, F. (1970b) 'The part played by labour in the transition from ape to man', in K. Marx and F. Engels, *Selected Works*, London, Lawrence & Wishart.

Enzensberger, H. (1974) 'A critique of political ecology', *New Left Review*, 84, March/April.

Esteva, G. (1986) 'Development is dangerous', *Resurgence*, 114, January/February.

Ewell, P. T. and Poleman, T. T. (1980) *Uxpanapa: Agricultural Development in the Mexican Tropics*, Oxford, Pergamon.

FAO (1980a) *Anuarrio de Produccion*, 34, Food and Agricultural Organization.

FAO (1980b) *Anuarrio de Fertilizantes*, 30, Food and Agricultural Organization.

FAO (1983) *Production Yearbook*, Rome, Food and Agricultural Organization.

FAO (1985) *Tropical Forestry Action Plan*, Rome, Food and Agricultural Organization.

Farvar, M. T. (1970) *The Careless Technology: Ecology and International Development*. New York, Natural History Press.

Farvar, M. T. and Glaeser, B. (1979) *Politics of Ecodevelopment*, Berlin, International Institute for Environment and Society.

Fearnside, P. M. (1985) 'Deforestation and decision-making in the development of Brazilian Amazonia', *Interciencia* 10(5): 243–7.

Foweraker, J. (1981) *The Struggle for Land*, Cambridge University Press.

Galtung, J. (1980) *Basic Needs and the Green Movement*, Tokyo, United Nations University.

Galtung, J. (1985) *Development Theory: Notes for an Alternative Approach*, Berlin, Wissenschaftszentrum, Reprints Series.

Geertz, C. (1971) *Agricultural Involution*, Berkeley, University of California Press.

George, S. (1985a) *Ill Fares the Land: Essays on Food, Hunger and Power*, London, Writers and Readers Publishing.

George, S. (1985b) *The Debt Crisis*, London, The Other Economic Summit.

Gibson, Charles (1981) *Los Aztecas Bajo el Dominio Espanol*, Mexico, Ed. Siglo XXI.

Gill, L. (1985) *Frontier Expansion and Rural Protest in Santa Cruz, Bolivia*, School of Economic and Social Studies, University of East Anglia, unpublished MS.

Glaeser, B. (1979) *Labour and Leisure in Conflict? Needs in Developing Industrial Societies*, Berlin, International Institute for Environment and Society (IIUG).

Gligo, N. (1985) *Avances en la interpretacion ambiental del desarollo agricola de America Latina*, Santiago, UNEP/ECLA.

Global 2000 (1982) *Report to the President*, London, Penguin.

Godau, R. (1985) 'La proteccion ambiental y la articulacion sociedad – naturaleza', *Estudios Sociologicos* 3(7), 47–84.

Godfrey, M. (1985) in T. Rose (ed.) *Crisis and Recovery in Sub-Saharan Africa*, Paris, OECD.

Gonzalez Pedrero, E. (1979) *La Riqueza de la Pobreza*, Mexico City, Joaquin Mortiz.

Goodland, R. (1981) 'Indonesia's environmental progress in economic development', in V. H. Sutlive *et al.* (eds.) *Deforestation in the Third World*, Williamsburg, Vermont.

Goodland, R. (1985) *Wildland Management in Economic Development*, Washington, DC, The World Bank.

Goodman, D. (1984) *Social Change in Brazil since 1945: Rural Economy and Society*, University College, London, Discussion Paper 85–103.

Goodman, D. and Redclift, M. (1981) *From Peasant to Proletarian: Capitalist Development and Agrarian Transitions*, Oxford, Blackwell.

Goodman, D. and Redclift, M. (1985), 'Capitalism, petty commodity production and the farm enterprise', *Sociologia Ruralis* 25(3/4): 231–47.

Goodman, D., Sorj, B. and Wilkinson, J. (1987) *From Farming to Biotechnology: The Industrialization of Agriculture*, Oxford, Blackwell.

Gorz, A. (1980) *Ecology as Politics*, London, Pluto Press.

Green, B. (1981) *Countryside Conservation*, London, Allen and Unwin.

Green, R. and Griffith-Jones, S. (1985) in T. Rose (ed.) *Crisis and Recovery in Sub-Saharan Africa*, Paris, OECD.

Gribben, J. (1979) *Future Worlds*, London, Abacus.

Grupo de Estudios Andres Ibanez (1983) *Tierra, estructura productiva y poder en Santa Cruz*, La Paz.

Hales, M. (1986) 'Management Science and "The Second Industrial Revolution" ', in Les Levidow (ed.) *Radical Science*, London, Free Association Books.

Hardin, G. (1968) 'The tragedy of the commons', *Science*, 162, 1243–8.

Harris, A. (1983) 'Radical economics and natural resources', *International Journal of Environmental Studies*, 21, 45–53.

Harrison, P. (1985) *Vicious Spirals*, a briefing paper for 'Seeds of Hope', London, Central Television.

Hayter, T. and Watson, C. (1985) *Aid: Rhetoric and Reality*, London, Pluto.

Hecht, S. (1985) *'Environment, Development and Politics: Capital Accumulation and the Livestock Sector in Eastern Amazonia'*, University of California, Los Angeles, unpublished MS.

Hecht, S., Anderson, A. and May, P. (1985) *The Subsidy from Nature: Successional Palm Forests and Rural Development*, Cornell University, Department of Rural Sociology, unpublished MS.

Hirsch, F. (1976) *The Social Limits to Growth*, London, Routledge and Kegan Paul.

Hodge, I. (1986) *Approaches to the Value of the Rural Environment*, paper presented to the Annual Conference of the Rural Economy and Society Study Group of the British Sociological Association, University of Loughborough.

Holling, C. S. (1978) (ed.) *Adaptive Environmental Assessment and Management*, Chichester, John Wiley.

Horton, D. E. (1984) *Social Scientists in Agricultural Research – Lessons from the Mantaro Valley Project, Peru*, IDRC (International Development Research Centre).

Hosier, R., O'Keefe, P., Wisner, B., Weiner, D. and Shakow, D. (1982) 'Energy Planning in Developing Countries', *Ambio*, 11(4).

Hubbard, L. J. (1986) 'The co-responsibility levy – a misnomer?', *Food Policy*, 11(3): 197–202.

Humphrey, C. R. and Buttel, F. (1982) *Environment, Energy and Society*, Belmont, California, Wadsworth.

Inglehart, R. (1981) 'Post-materialism in an environment of insecurity', *American Political Science Review*, 75, 279–316.

Ingram, B. (1983) 'Parks in the twenty-first century', *Not Man Apart*, March.

IUCN (1980) *World Conservation Strategy*, Gland, Switzerland, International Union for the Conservation of Nature.

Jackson, T. and Eade, D. (1982) *Against the Grain*, Oxford, Oxfam.

de Janvry, A (1981) *The Agrarian Question and Reformism in Latin America*, Baltimore, Johns Hopkins University.

Jeans, D. N. (1983) 'Wilderness, nature and society: contributions to the history of an environmental attitude', *Australian Geographical Studies* 21, October, 170–82.

Johnson, B. (1986) 'Outside in', *Guardian*, 31 October.

Jorgensen, H. (1973) 'Problems of tropical settlements – experiences in Colombia and Bolivia' in M. T. Farvar and J. P. Milton (eds.) *The Careless Technology*, London, Stacey.

Katzman, M. T. (1975) 'The Brazilian frontier in comparative perspective', *Comparative Studies in Society and History*, 17(3), July.

Kearney, M. (1986) 'Integration of the Mixteca and the Western US–Mexico Border Region via migratory labour', in I. Rosenthal-Urey (ed.), *Regional Impacts of US–Mexican Relations*, San Diego, Center for US–Mexican Studies, University of California, Monograph 16.

Kenney, M. and Buttel, F. (1985) 'Biotechnology and international development: prospects for overcoming dependence in the information age', *Rural Sociology Bulletin*, 143, Cornell University.

Khozin, G. (1979) *The Biosphere and Politics*, Moscow, Progress Publishers.

Kuhn, T. S. (1962) *The Structure of Scientific Revolution*, Chicago, University of Chicago Press.

Lappé, F., Collins, J. and Kinley, D. (1981) *Aid as Obstacle*, San Francisco, Institute for Food and Development Policy.

Lashof, D. (1986) *The Ecology of Socialism*, Energy and Resources Group, University of California, Berkeley, unpublished MS.

Lawrence-Jones, W. (1984) *Comentarios de la implementacion de un sondeo en los Valles Mesotermicas*, Santa Cruz, Bolivia, British Technical Agricultural Mission (BTAM) Working paper no. 37.

Lecomte, B. (1985) in T. Rose (ed.) *Crisis and Recovery in Sub-Saharan Africa*, Paris, OECD.

Leder, P. (1982) 'Moving genes: promises kept and pending', *Miami Winter Symposium*, Miami, Florida.

Lefever, L. and North, L. (1980) 'Introduction' to Lefever, L. and North, L. (eds.) *Democracy and Development in Latin America*, Toronto, York University.

Leff, E. (1985) 'Ethobotanics and anthropology as tools for a cultural conservation strategy', in J. McNeely and D. Pitt (eds.) *Culture and Conservation: The Human Dimension in Environmental Planning*, London, Croom Helm.

Lenin, V. I. (1972) *Imperialism, The Highest Stage of Capitalism*, Moscow, Progress Publishers.

Lipset, S. M. and Solari, A. (eds.) (1967) *Élites in Latin America*, Oxford University Press.

Lipton, M. (1977) *Why Poor People Stay Poor*, London, Temple Smith.

Lopez Cordovez, L. (1982) 'Trends and recent changes in the Latin American food and agriculture situation', *CEPAL Review* 16, Santiago, Chile, United Nations Economic Commission for Latin America.

Lowe, P. and Rudig, W. (1984) 'Political ecology and the social sciences – the state of the art', *British Journal of Political Science*, 16, 513–50.

Luiselli, C. (1985) *The Route to Food Self-Sufficiency in Mexico: interactions with the US food system*, San Diego, Center for US–Mexican Studies, University of California.

Luxemburg, R. (1951) *The Accumulation of Capital*, London, Routledge and Kegan Paul.

Maathai, W. (1985) 'The Green Belt Movement: building blocks for sustainable development', paper to 'The Other Economic Summit', London.

Maathai, W. (1986) 'Learning from the South' in P. Elkins (ed.) *The Living Economy*, London, Routledge and Kegan Paul.

MacFadyen, J. (1985) 'A battle over seeds', *Atlantic*, November.

Mackillop, A. (1980) 'Energy for the developing world', *Energy Policy*, December.

McNeely, J. and Pitt, D. (eds.) (1985) *Culture and Conservation: The Human Dimension in Environmental Planning*, London, Croom Helm.

Martinez-Alier, J. (1985) *Modern Agriculture: a Source of Energy?*, Barcelona, unpublished MS.

Maslow, A. H. (1954) *Motivation and Personality*, New York, Harper & Row.

Marx, K. (1974) *Capital*, volume III, fifth edition, London, Lawrence and Wishart.

Maxwell, S. (1979) *Colonos Marginalizados al Norte de Santa Cruz – Avenidos de Escape de la Crisis del Barbecho*, September, CIAT/BTAM Working Paper 4.

Meadows, D. H., Meadows, D. L., Randers, J. and Behrens, W. (1972) *The Limits to Growth*, London, Pan.

Meillassoux, C. (1981) *Maidens, Meal and Money: Capitalists and the Domestic Community*, Cambridge University Press.

Mooney, P. R. (1979) *Seeds of the Earth*, London, ICDA.

Morales, Hector Luis (1984) 'Chinampas and integrated farms: learning from rural traditional experience', in F. Di Castri and M. Hadley, *Ecology in Practice*, Paris, UNESCO.

Moran, E. (1984) 'Amazon Basin colonization' *Interciencia*, 9(6).

Myers, N. (1979) *The Sinking Ark*, Oxford, Pergamon.

Myers, N. (1985) *Population, Environment and Conflict*, paper for UNFPA Conference on Population, Development and Peace, London 15–17 May.

Myers, N. and Myers, D. (1982) 'Increasing awareness of the supranational nature of emerging environmental issues', *Ambio*, xi, 4.

Naess, A. (1973) 'The shallow and the deep, long-range ecology movement. A summary', *Inquiry*, 16, 95–100.

Naess, A. (1983) *Philosophical Aspects of the Deep Ecological Movement*, Oslo, unpublished MS.

Nations, B. and Korner, L. (1983) 'Cattle eat the forest', *Environment*, 25(2).

Norgaard, R. (1981) 'Sociosystem and ecosystem coevolution in the Amazon', *Journal of Environmental Economics and Management*, 8, 238–54.

Norgaard, R. (1984a) 'Coevolutionary agricultural development', *Economic Development and Cultural Change*, 32(3).

Norgaard, R. (1984b) 'Coevolutionary development potential', *Land Economics*, 60(2).

Norgaard, R. (1985a) 'Environmental economics: an evolutionary critique and a plea for pluralism', *Journal of Environmental Economics and Management*, 12(4).

Norgaard, R. (1985b) *The Scarcity of Resource Economics*, paper presented to the American Economics Association, New York.

Nossal, G. J. (1985) *Reshaping Life: Key Issues in Genetic Engineering*, Cambridge University Press.

Nowicki, P. (1985) 'Cultural ecology and "management" of natural resources or knowing when not to meddle', in J. McNeely and D. Pitt, *Culture and Conservation: The Human Dimension in Environmental Planning*, London, Croom Helm.

O'Brien, P. (1986) 'The debt cannot be paid: Castro and the Latin American debt', *Bulletin of Latin American Research*, 5(1).

Odum, H. E. (1971) *Environment, Power and Society*, New York, John Wiley.

Ojeda, O. and Sanchez, V. (1985) 'La cuestion ambiental y la articulacion sociedad – naturaleza', *Estudios Sociologicos* 3(7), 25–46.

Oliveira, F. de (1972) 'A economia brasileira: critica da razao dualista', *Estudos CEBRAP*, 2.

O'Riordan, T. (1981) *Environmentalism*, London, Pion.

O'Riordan, T. (1986) *Sustainability and the wider environmental movement*, paper given to the Economic and Social Research Council Working Group on Environmental Economics, University of East Anglia.

Ovington, J. and Fox, A. (1980) *Wilderness – A Natural Asset*, II International

Congress on Wilderness, Cairns, Australia.

Park, Robert E. and Burgess, Ernest W. (1921) *Introduction to the Science of Sociology*, Chicago, University of Chicago Press.

Pearce, D. (1985) *Sustainable Futures: economics and the environment*, inaugural lecture, Department of Economics, University College, London, 5 December.

Pearce, D. (1986) *The Sustainable Use of Natural Resources in Developing Countries*, paper to the Economic and Social Research Council Workshop on Environmental Economics, University of East Anglia.

Pepper, D. (1984) *The Roots of Modern Environmentalism*, London, Croom Helm.

Perry, J. (1986) 'Managing the world environment', *Environment*, 28(1).

Plumwood, V. and Routley, R. (1982) 'World rainforest destruction: the social factors', *The Ecologist*, 12(1).

Prebisch, R. (1976) 'Critica al capitalismo periferico', *Revistade CEPAL*, 1.

PRUSDA (1984) Comision Coordinadora Para el Desarrollo Agropecuario de Distrito Federal, Mexico.

Rama, R. and Vigorito, R. (1979) *El Complejo de Frutas y Legumbres en Mexico*, Mexico, Nueva Imagen.

Redclift, M. R. (1984) *Development and the Environmental Crisis: Red or Green Alternatives?* London, Methuen.

Rich, B. (1985) 'Multilateral development banks – their role in destroying the global environment', *The Ecologist*, 15, 1/2.

Richards, P. (1985) *Indigenous Agricultural Revolution – Ecology and Food Production in West Africa*, London, Hutchinson.

Riddell, R. (1981) *Ecodevelopment*, London, Gower.

Rifkin, J. (1985) *Declaration of a Heretic*, London, Routledge and Kegan Paul.

Rivera, R. (1985) *The Role of Temporary Rural Work in Chile under the Neo-Liberal Development Policy*, Ph.D. thesis, University of Durham.

Rose, S., Lewontin, R. and Kamin, L. (1984) *Not In Our Genes: Biology, Ideology and Human Nature*, London, Penguin.

Rosenbaum, W. A. (1973) *The Politics of Environmental Concern*, New York, Praeger.

Roxborough, I. (1979) *Theories of Underdevelopment*, London, Macmillan.

Russell, P. (1982) *The Awakening Earth: Our Next Evolutionary Leap*, London, Routledge & Kegan Paul.

Sachs, I. (1976) 'Environment and styles of development', in W. H. Matthews (ed.) *Outer Limits and Human Needs*, Uppsala.

Sachs, I. (1980) *Strategies de l'écodeveloppement*, Paris, Editions Ouvrières.

Sachs, I. (1984) 'The strategies of ecodevelopment', *Ceres*, 17(4).

Sage, C. (1985) *Risk maximization and livelihood: potato producers in the Cochabamba Serranía*, Bolivia. Paper delivered to Symposium on 'Environment and Society in the Third World', 45th Congress of Americanists, Bogota, Colombia.

Saint, W. (1982) 'Farming for energy: social options under Brazil's national alcohol programme', *World Development*, 10(3).

Sanchez de Carmona, L. (1984) 'Ecological studies for regional planning in the Valley of Mexico', in F. Di Castri and M. Hadley (eds.) *Ecology in Practice*, Paris, UNESCO.

Sandbach, F. (1980) *Environment, Ideology and Policy*, Oxford, Blackwell.

Sandbrook, R. (1982) *The Conservation and Development Programme for the UK: a response to the World Conservation Strategy*, (Chapter 5), World Wildlife Fund.

Sanders, J. and Lynam, J. (1981) 'New agricultural technology and small farmers in Latin America', *Food Policy*, 6(1).

Sanderson, S. (1986) *The Transformation of Mexican Agriculture*, Princeton University Press.

Sawyer, D. R. (1979) *Peasants and Capitalism on the Amazon Frontier*, Harvard University Ph.D. thesis.

Sayer, A. (1983) 'Notes on geography and the relationship between people and nature', in *Society and Nature*, London Group of the Union of Socialist Geographers.

Schmidt, A. (1971) *The Concept of Nature in Marx*, London, New Left Books.

Schuurman, F. J. (1979) 'Colonization policy and peasant economy in the Amazon basin', *Boletin de Estudios Latinoamericanos y del Caribe*, 27 December.

Shiva, V. (1986) 'Coming tragedy of the commons', *Economic and Political Weekly*, 21(15), 12 April.

Shiva, V. and Bandyopadhyay J. (1986) 'The evolution, structure and impact of the Chipko movement', *Mountain Research and Development*, 6(2), May.

Shoard, M. (1980) *The Theft of the Countryside*, London, Temple Smith.

de Silva, S. B. D. (1982) *The Political Economy of Underdevelopment*, London, Routledge & Kegan Paul.

Simmons, I. (1974) *The Ecology of Natural Resources*, London, Edward Arnold.

Singer, L. and Ansari, S. (1977) 'Food aid and food credibility', *Alternative Medicine*, 3(1).

Skinner, B. (1985) 'Cattle ranching in Brazil', *The Ecologist*, 14(2).

Slater, D. (ed.) (1985) *New Social Movements and the State in Latin America*, Amsterdam, CEDLA.

Smith, N. (1984) *Uneven Development*, Oxford, Blackwell.

Smith, N. and O'Keefe, P. (1980) 'Geography, Marx and the concept of Nature', *Antipode*, 12(2).

Sohn-Rethel, A. (1986) 'Science as alienated consciousness' in Les Levidow (ed.) *Radical Science Essays*, London, Free Association Books.

Sorj, B. and Pompmermayer, M. J. (1983) *Sociedade e Politica(s) na Fronteira Amazonica: Interpretacoes e Argumentos*, Belo Horizonte, unpublished MS.

South (magazine) (1985) 'Latin America 600', January 63–81.

South Pacific Commission (1980) *Comprehensive Environmental Management Plan*, Nonanla, New Caledonia.

Spalding, K. (1984) *The Mexican Food Crisis: an analysis of the SAM*, San Diego, Center for US–Mexican Studies, Research Report 33.

Stanley, D. L. (1984) *Organisations in the Sierra Region of Santo Domingo: associated credit and community development*, unpublished MS, St Antony's College, Oxford.

Starlinger, P. (1982) in W. J. Whelan and S. Black (eds.) *From Genetic Engineering to Biotechnology – the Critical Transition*, London, Wiley.

Sunkel, O. and Gligo, N. (1980) *Estilos de Desarrollo, Energia y Medioambiente*, Santiago, CEPAL.

Sylvan, R. (1985a and b) 'A critique of Deep Ecology', *Radical Philosophy*, 40 and 41.

Taussig, M. (1980) *The Devil and Commodity Fetishism in South America*, Chapel Hill, University of North Carolina Press.

Taylor, P. L. (1983) *The San Julian Multipurpose Cooperative*, FIDES, Santa Cruz, mimeo.

Thiede, W. (1984) '10 Jahre versorgungsberechnungen für die EG' (10 years of supplies to the EEC), *Agrarwirtschaft*, 5.

Thompson, M. (1986) *The Cultural Construction of Nature and the Natural Destruction of Culture*, Center for Philosophy and Public Policy, University of Maryland, Working Paper 8.

Thompson, M., James, P. and Taylor, P. (1986) *Environmental Rationalities*, Institute for Management Research, University of Warwick MS.

Thomson, G. (1978) *The First Philosophers*, London, Lawrence and Wishart.

Timberlake, L. (1985) *Africa in Crisis*, London, Earthscan.

Timothy, D. (1973) 'Plant germ plasm resources and utilization', in

216 SUSTAINABLE DEVELOPMENT

M. T. Farvar and J. P. Milton (eds.) *The Careless Technology*, London, Stacey.

Tinker, J. (1983) 'Fending off disaster', *People* (IPPF), 10(1).

Tobias, M. (ed.) (1984) *Deep Ecology*, San Diego, Avant Books.

Toledo, A. (1985) *Como Destruir el Paraiso*, Mexico City, Centro de Ecodesarollo.

Toledo, V., Carabias, J., Mapes, C. and Toledo, C. (1981) 'Critica de la ecologia politica', *Nexos*, 47.

Trainer, F. E. (1986) *Abandon Affluence!*, London, Zed Books.

UN (1981) *World Population*, New York, United Nations.

UN (1982) *Technology, Trade and Transnational Corporations in the Food Processing Sector of Mexico: A Case Study*, Geneva, Committee on Transfer
of Technology.

UN (1977) *Desertification: An Overview*, Nairobi, United Nations Conference on Desertification.

UNEP (1975) *The Proposed Programme*, Nairobi, United Nations Environment Programme.

UNEP (1981) *In Defence of the Earth*, Nairobi, United Nations Environment Programme.

UNEP (1984) *General Assessment of Progress in the Implementation of the Plan of Action to Combat Desertification*, Nairobi, United Nations Environment Programme.

Velho, O. (1976) *Capitalismo, autoritario e campesinato*, São Paulo, DIFEL.

Veliz, C. (1967) *The Politics of Conformity in Latin America*, Oxford University Press.

Vitale, L. (1983) *Hacia Una Historia Del Ambiente en America Latina*, Mexico City, Nueva Imagen.

Watson, C. (1985) 'Working at the World Bank' in T. Hayter and C. Watson (1985) *Aid: Rhetoric and Reality*, London, Pluto.

Weir, D. and Schapiro, M. (1981) *Circle of Poison*, San Francisco, Institute for Food and Development Policy.

Whiteford, S. (1986) 'Troubled waters: the regional impact of foreign investment and state capital in the Mexicali valley', in I. Rosenthal-Urey (ed.) *Regional Impacts of US–Mexican Relations*, San Diego, Center for US–Mexican Studies, University of California, Monograph 16.

Wiest, R. (1984) *Mixtec migration from Oaxaca to northwest Mexico and California*, paper presented to Conference on Regional aspects of US–Mexican Integration, University of California, San Diego, California.

Wilkes, G. (1983) 'Current status of crop plant germplasm', *Critical Reviews in Plant Science*, 1(2).

Williams, R. (1984) *Towards 2000*, London, Penguin.

Wilson, E. O. (1975) *Sociobiology: the New Synthesis*, Cambridge, Mass., Harvard University Press.

Wilson, E. O. (1978) *On Human Nature*, Cambridge, Mass., Harvard University Press.

Wittfogel, K. (1981) *Oriental Despotism*, New York, Vintage Books.

Wolf, E. (1982) *Europe and the People without History*, Berkeley, University of California Press.

Wolf, E. C. (1985) 'Challenges and priorities in conserving biological diversity', *Interciencia*, 10(5). 236–42.

Wolf, W. (1959) *Sons of the Shaking Earth*, University of Chicago Press.

World Bank (1979) *Environment and Development*, Washington DC.

WCED (1985) World Commission for Environment and Development.

WCS (1983) *World Conservation Strategy*, Living Resource Conservation for Sustainable Development, Gland, Switzerland, International Union for Conservation of Nature and Natural Resource.

Yates, P. L. (1981) *Mexico's Agricultural Dilemma*, Tucson, University of Arizona Press.

Index